T0210556

Lecture Notes in Computer Science 9132

Commenced Publication in 1973
Founding and Former Series Editors:
Gerhard Goos, Juris Hartmanis, and Jan van Leeuwen

More information about this series at http://www.springer.com/series/7412

Friedhelm Schwenker · Fabio Roli
Josef Kittler (Eds.)

Multiple Classifier Systems

12th International Workshop, MCS 2015
Günzburg, Germany, June 29 – July 1, 2015
Proceedings

 Springer

Editors
Friedhelm Schwenker
Ulm University
Ulm
Germany

Fabio Roli
University of Cagliari
Cagliari
Italy

Josef Kittler
University of Surrey
Guildford
UK

ISSN 0302-9743   ISSN 1611-3349   (electronic)
Lecture Notes in Computer Science
ISBN 978-3-319-20247-1   ISBN 978-3-319-20248-8   (eBook)
DOI 10.1007/978-3-319-20248-8

Library of Congress Control Number: 2015940974

LNCS Sublibrary: SL6 – Image Processing, Computer Vision, Pattern Recognition, and Graphics

Printed on acid-free paper

Springer International Publishing AG Switzerland is part of Springer Science+Business Media
(www.springer.com)

Preface

This book presents the proceedings of the 12th IAPR Workshop on Multiple Classifier Systems (MCS 2015) held at Reisensburg Castle, the research center of Ulm University, Germany, during June 29 – July 1, 2015. The series of MCS workshops has acted as a major forum for international researchers and practitioners from the community of multiple classifier systems in pattern recognition and machine learning.

The Program Committee of MCS 2015 selected 19 papers for the scientific program. Two IAPR Invited Sessions given by Dr. George Cybenko, Darmouth College, USA, and Dr. Marcello Pelillo, University of Venice, Italy, enriched the workshop.

MCS 2015 would not have been possible without the help and support of many people and organizations. First of all, we are grateful to all authors who submitted their work to MCS. We thank the members of the Program Committee and the additional reviewers for performing the difficult task of selecting the best papers for presentation, and we hope that readers of this volume will enjoy it and be inspired from its contributions.

MCS 2015 was supported by the International Association for Pattern Recognition (IAPR), by the University of Cagliari, Italy, by the University of Surrey, UK, and by Ulm University, Germany, which hosted this event. Special thanks to the people of the local organization, Miriam Schmidt, Martin Schels, Michael Glodek, Markus Kächele, Sascha Meudt, and Patrick Thiam. Finally, we wish to express our gratitude to Springer for publishing our proceedings in their LNCS series and for their constant support.

July 2015

Friedhelm Schwenker
Fabio Roli
Josef Kittler

Organization

Organizing Committee

Friedhelm Schwenker   University of Ulm, Germany
Fabio Roli   University of Cagliari, Italy
Josef Kittler   University of Surrey, UK

Program Committee

Jon Benediktsson, Iceland
Gavin Brown, UK
Cesare Furlanello, Italy
Giorgio Fumera, Italy
Joydeep Ghosh, USA
Larry Hall, USA
Tin Kam Ho, USA
Philip Kegelmeyer, USA
Ludmila Kuncheva, UK

Gonzalo Martinez-Munoz, Spain
Juan Jose Rodriguez, Spain
Arun Ross, USA
Carlo Sansone, Italy
Giorgio Valentini, Italy
Terry Windeatt, UK
Xin Yao, China
Yang Yu, China
Zhi-Hua Zhou, China

Sponsoring Institutions

University of Ulm, Germany
University of Cagliari, Italy
University of Surrey, UK
International Association for Pattern Recognition (IAPR)

Deep Learning of Behaviors

George Cybenko

Thayer School of Engineering
Dartmouth College
Hanover NH 03755, USA
gvc@dartmouth.edu

Abstract. Deep learning has generated much research and commercialization interest recently. In a way, it is the third incarnation of neural networks as pattern classifiers, using insightful algorithms and architectures that act as unsupervised auto-encoders, learning hierarchies of features in a dataset.

After a short review of that work, we will discuss the challenges associated with the analysis of behaviors observed as time series of categorical data. Novel computational approaches for deep learning of behaviors as opposed to just static patterns will be presented. Our approach is based on structured non-negative matrix factorizations of matrices that encode observation frequencies of behaviors.

These techniques can be used to robustly characterize and exploit diverse behaviors in security applications such as covert channel detection and coding. Examples of such applications will be presented.

Related results about the role of diversity in computer security applications will also be introduced wherein adversarial dynamics dictates that attackers and defenders coevolve. As a result, the use of multiple diverse detection and mitigation techniques makes the attackers' effective workfactor much higher.

Similarity-Based Pattern Recognition:

A Game-Theoretic Perspective

Marcello Pelillo

Ca' Foscari University of Venice
30172 Venezia Mestre, Italy
pelillo@dsi.unive.it

Abstract. Similarity-based methods are emerging as a powerful tool in pattern recognition and machine learning because of their ability to overcome the intrinsic limitations of traditional feature-vector approaches. By departing from vectorspace representations, however, one is confronted with the challenging problem of dealing with (dis)similarities that do not necessarily possess the Euclidean behavior or not even obey the requirements of a metric. In this talk, I will maintain that game theory offers an elegant and powerful conceptual framework which serves well this purpose, and I will describe recent attempts aimed at formulating various similarity-based pattern recognition problems from a game-theoretic perspective. Particular emphasis will be given to evolutionary-based models which, in contrast to the classical theory, offer an intriguing dynamical system perspective. Finally, I will descrive some applications of this approach within the context of multiple classifier systems.

Contents

Application and Evaluation

Theory and Algorithms

A Novel Bagging Ensemble Approach for Variable Ranking and Selection for Linear Regression Models

Chun-Xia Zhang[1]([✉]), Jiang-She Zhang[1], and Guan-Wei Wang[2]

[1] School of Mathematics and Statistics, Xi'an Jiaotong University,
Xi'an 710049, Shaanxi, China
{cxzhang,jszhang}@mail.xjtu.edu.cn
[2] School of Mechatronic Engineering, Xi'an Technological University,
Xi'an 710021, Shaanxi, China
waldowgw@163.com

Abstract. With respect to variable selection for linear regression models, a novel bagging ensemble method is developed in this paper based on a ranked list of variables. Specifically, a mixed importance measure is assigned to each variable according to the order that it is selected by stepwise search algorithm into the final model as well as the improvement resulted from its inclusion. Considering that small permutations in training data may lead to some changes in the order that the variables enter the final model, the above process is repeated for multiple times with each executed on a bootstrap sample. Finally, the importance measure of each variable is averaged across the bootstrapping trials. The experiments conducted with some simulated data demonstrate that the novel method compares favorably with some other variable selection techniques.

Keywords: Variable selection · Variable ranking · Bagging · Stepwise search algorithm · Parallel genetic algorithm · Ensemble learning

1 Introduction

Given some observations of $(y, \mathbf{X})$, linear regression model is a basic model to explore how the covariates $X_1, X_2, \cdots, X_p$ influence the response variable Y due to its simplicity and effectiveness. In statistical learning field, there are two fundamental goals: ensuring high prediction accuracy and discovering truly informative variables. Because variable selection can result in better prediction accuracy as well as a concise model for interpretation purpose, it has always been an important topic in linear regression analysis. Nowadays, with the vast availability of high-dimensional data in various real applications, variable selection has gained more interest of researchers to cope with related problems [1–4]. Here, it is noteworthy that predictive modelling and interpretation modelling are quite different as indicated by many researchers [2,5,6], and we will deal with the second goal in the current work.

© Springer International Publishing Switzerland 2015
F. Schwenker et al. (Eds.): MCS 2015, LNCS 9132, pp. 3–14, 2015.
DOI: 10.1007/978-3-319-20248-8_1

At present, there exist a large variety of variable selection techniques in the literature, such as subset selection [7,8], coefficient shrinkage [3,9–13] and so on. However, variable selection has one disadvantage that different sizes of model require different tuning parameters in the analysis, which is hard to specify even for statisticians. In view of this point, some researchers [6] advocated variable ranking instead of variable selection since the selection can be performed by adopting a thresholding rule once the variables are ranked properly. In this paper, we will discuss some methods to achieve variable ranking and selection for linear regression models. The selection will be implemented on the basis of a ranked list of variables.

In recent years, ensemble methods have become popular in the variable selection context, for example, parallel genetic algorithm (PGA) [14], stability selection [10,11], random lasso [15] and stochastic stepwise ensembles (ST2E) [6]. Ensemble learning [16], a relatively new paradigm in machine learning field, generally utilizes many learning machines to solve a problem so that these machines can complement each other. Up to now, ensemble learning techniques have make great success in various domains, particularly for the prediction problems (i.e., classification and regression) that are often encountered in practice [17–20]. Here, we must differentiate the term "prediction ensemble (PE)" from "variable-selection ensemble (VSE)" since they are used for quite different purposes. Specifically, PEs aims to maximize *prediction accuracy* so that future data can be well predicted. As far as VSEs are concerned, the purpose is to maximize *selection accuracy* in order that the underlying important variables can be identified as accurate as possible.

To achieve high selection accuracy, we propose in this paper a novel ensemble approach to build VSEs. For ease of presentation, the novel method will be abbreviated as BSSW throughout this paper since it is motivated by applying bagging technique to stepwise search algorithm. The main idea of BSSW is summarized as follows. In order to produce multiple slightly different measures of importance for each variable, we draw some bootstrap samples and apply stepwise search algorithm to each sample. Based on the order that one variable is selected into the final model as well as the improvement of model fitting resulted from its inclusion, a weight is assigned to it with the purpose to reflect its importance to the response variable. Finally, the importance measure of each variable is averaged across the bootstrapping trials. All the variables are then ranked in the light of this averaged importance measure. If the final goal is to select significant variables, we can search for the largest gap between any consecutive entries and choose the variables that are located above the gap. Based on the experiments conducted with some simulated data, the newly proposed method BSSW is shown to compare favorably with some other variable selection techniques.

The rest of this paper is organized as follows. Section 2 gives a brief introduction of VSEs. In Sect. 3, the novel ensemble approach BSSW is described in detail to implement variable ranking and selection for linear regression models. This is followed by conducting some experiments to examine and compare the performance of BSSW with some other procedures in Sect. 4. Finally, Sect. 5 offers the conclusions of the paper.

2   Brief Introduction of VSEs

Suppose that there are p variables, a VSE (of size B) can be represented by a $B \times p$ matrix, say, $\mathbf{E}$, whose jth column contains B repeated but slightly different measures of how important the jth variable is. On the basis of $\mathbf{E}$, one can obtain the average importance of each variable using a majority voting rule, that is,

$$R(j) = \frac{1}{B} \sum_{b=1}^{B} \mathbf{E}(b, j), \quad j = 1, 2, \cdots, p, \tag{1}$$

where $\mathbf{E}(b, j)$ stands for the (b, j)th entry of $\mathbf{E}$. Subsequently, the p variables can be ordered and those variables ranked "considerably higher" than the rest are then selected. To make selection decision, one can select variable j if it is ranked above average, that is, if $R(j) > (1/p) \sum_{k=1}^{p} R(k)$. In addition, one can also look for an elbow from the so-called scree plot, which is a very common practice in cluster analysis. Specifically, it can be done as follows: sort the values $R(1), R(2), \cdots, R(p)$ in a descending order; search for the biggest gap between any consecutive entries; and select the variables that are located above the gap. In all the experiments reported later, the latter strategy will be adopted.

The key to construct a good VSE lies in producing multiple measures of importance for each candidate variable. In contrast, traditional variable selection methods such as subset selection and lasso only produce one measure, that is, $B = 1$. Generally speaking, averaging over a number of independent measures is often beneficial. The great success of prediction ensembles in many areas has provided a vivid example of this issue. The situation is similar for VSEs. This is also the main reason why VSEs are attractive and more powerful than many traditional procedures.

Analogous to the techniques for creating PEs, there can be many ways to construct VSEs. But currently, the commonly used method is to implement a stochastic mechanism so that we can repeatedly perform traditional variable selection and obtain slightly different answers each time. One way is to use a stochastic rather than deterministic search algorithm to perform variable selection, for example, PGA [14], RandGA [21] and ST2E [6]. Another way to build a VSE is to perform the selection on bootstrap samples. Stability selection [10,11], random lasso [15] and bagged stepwise search (BSS) [22] belong to this type of methods. Due to space constraint of this paper, the details of these methods will be omitted here.

3   Novel Ensemble Approach BSSW for Variable Ranking and Selection

As for the traditional variable selection methods like subset selection and stepwise search algorithm, they usually produce one importance measure for each variable. Under some circumstances, a small permutation in the data may lead them to select quite different variables. Put in another way, the variation of data makes these methods easily exclude some truly important variables or falsely pick some unimportant variables. Evidently, this largely prevents the use of these methods in real-world applications. Take stepwise selection as an example, it may suffer from high variability as pointed out in [13]. Meanwhile, it ignores the stochastic errors or uncertainty in the variable selection stage. With the aim to exploit stepwise search to accurately detect the variables that are truly influential to the response, we have to first overcome its above shortcomings.

Due to the existence of noise in data, some spurious variables may be falsely considered as important ones by chance. However, the truly important ones will be often included into the final model. More specifically, if the selection process is conducted using slightly different data for a number of trials $b = 1, 2, \cdots, B$, the measure $\mathbf{E}(b, j)$ for an actually influential variable j will be high for all or most of b. Nevertheless, $\mathbf{E}(b, j)$ will be high only for some b if a variable j is spurious. Therefore, when averaged over B trails, the importance measure $R(j)$ will be high for the variables that are truly important. This explains to a large extent why VSEs gain significant interest in the variable selection context.

Note that the output of some traditional variable selection methods also includes the order that each variable is selected into the model besides the chosen variables. As far as we know, however, this information has not been utilized in the existing techniques for generating VSEs. On the other hand, the characteristics of stepwise selection (i.e., high variability, often trapped into a local solution and etc.) make it appropriate to behave as the base learner to build a VSE. To make full use of the information included in its output, we propose a novel bagging ensemble approach to implement variable ranking and selection via defining a mixed importance measure for each variable.

To produce a series of slightly different importance measures for each candidate variable, we first draw some bootstrap samples from the given data set. Subsequently, stepwise search algorithm is applied to each sample. On the basis of the results acquired from a bootstrap sample, we can find out whether one variable is selected or not. If a variable occurs in the final model, the order that it enters the model can also be obtained. According to this information and the improvement of model fitting resulted from its inclusion, we assign a weight to each variable in order to reflect its importance to the response variable. In the meantime, the weights associated with all the variables are normalized so that their sum is equal to 1. Finally, the importance measure of each variable is averaged across the bootstrap trials. The variables can then be ranked in the light of this averaged importance measure. If the purpose is just to analyze the relative importance of the exploratory variables on the response, the ranked results can meet the requirements and the algorithm can stop here. Otherwise, a thresholding rule or some other technique (such as making a scree plot and looking for an elbow) needs to be adopted so that important variables for interpretation can be identified. The main steps of utilizing BSSW to achieve variable ranking and selection are listed in Fig. 1.

In step (c) in the proposed algorithm, the main reason why we allocate a weight as defined in (2) to each variable is explained as follows. Notice that after executing stepwise search on a bootstrap sample according to a certain criterion, its output

Input

$\mathscr{L}$: a given training set $\mathscr{L} = \{(\mathbf{x}_i, y_i)\}_{i=1}^N$, where $\mathbf{x}_i \in \mathbb{R}^p$ and $y_i \in \mathbb{R}$.

B: ensemble size.

Output

Variable ranking: a ranked list of the p variables;

Variable selection: the selected variables.

Main process of BSSW

1. Initialize the elements of $\mathbf{E}$ to be 0.
2. For $b = 1, 2, \cdots, B$
   (a) Randomly draw N training instances from $\mathscr{L}$ with replacement to compose a new training set $\mathscr{L}_b = \{(\mathbf{x}_i^{(b)}, y_i^{(b)})\}_{i=1}^N$.
   (b) Provide $\mathscr{L}_b$ as the input of stepwise search algorithm to perform variable selection according to a criterion (e.g., AIC, BIC, C_p).
   (c) For the variables included into the final model, assign a weight to each of them, that is,

$$\mathbf{E}(b, j) = rank(j) + \ln|improve(j)|, \quad j = 1, 2, \cdots, p, \qquad (2)$$

   where "$rank(j)$" is an importance value assigned to variable j according to the order that it is selected into the model. The symbol $improve(j)$ indicates how much variable j improves the model fitting due to its inclusion.
   (d) Normalize the weights assigned to each variable, namely,

$$\widetilde{\mathbf{E}}(b, j) = \frac{\mathbf{E}(b, j)}{\sum_{k=1}^p \mathbf{E}(b, k)}, \quad j = 1, 2, \cdots, p. \qquad (3)$$

3. EndFor
4. Calculate the averaged importance measure for each variable as

$$R(j) = \frac{1}{B} \sum_{b=1}^B \widetilde{\mathbf{E}}(b, j), \quad j = 1, 2, \cdots, p. \qquad (4)$$

5. According to $R(1), R(2), \cdots, R(p)$, sort the variables in descending order. If the target is variable ranking, stop the algorithm here. Otherwise, execute the following step.
6. By adopting a further selection step, to select the variables that are most influential to the response variable.

Fig. 1. The novel ensemble approach BSSW for variable ranking and selection.

also embodies the order that each variable is selected into the model except for the chosen variables and their coefficient estimation. The earlier one variable is chosen, the bigger influence it has on the response. On the other hand, the inclusion of each variable will lead to some improvement of the fitted model. Generally speaking, the larger the improvement of one variable is, the more important it is. Thus, we combine these two terms to assign a mixed importance measure as defined in (2) to each variable. In the formula (2), the earlier a variable j is selected, the larger the value of $rank(j)$ is. When computing this term in one experiment, we let the first selected variable receive value p, the second one receive value $p-1$, $\cdots$ and the last one have value 1. As for $improve(j)$, it can be calculated as the decrement of AIC (or BIC) value after adding a new variable into the model.

4   Experimental Study

In this section, we will carry out experiments with some simulated data to investigate the performance of the proposed method BSSW. Meanwhile, BSSW will be also compared with some other variable selection techniques including single-path genetic algorithm (SGA) [14], parallel genetic algorithm (PGA) [14], traditional stepwise search algorithm (Stepwise) [8], bagged stepwise search (BSS) [22].

4.1   Simulation 1

Similar to the simulated data used in [14], in this experiment we randomly generated a data set consisting of $n = 40$ observations and $p = 20$ variables. The variables are generated from normal distributions and the model is

$$\mathbf{y} = \mathbf{x}_5 + 2\mathbf{x}_{10} + 3\mathbf{x}_{15} + \epsilon, \quad \epsilon \sim N(\mathbf{0}, \sigma^2 \mathbf{I}), \tag{5}$$

where ϵ is an error term. Obviously, only variables 5, 10 and 15 have actual influence on the response y. Regarding the mean and covariance for the variables, we considered the following 5 variations:

Variation 0: $\mathbf{x}_1, \mathbf{x}_2, \cdots, \mathbf{x}_{20} \sim N(\mathbf{0}, \mathbf{I})$;
Variation 1: $\mathbf{x}_1, \mathbf{x}_2, \cdots, \mathbf{x}_{19} \sim N(\mathbf{0}, \mathbf{I})$, $\mathbf{x}_{20} = \mathbf{x}_5 + 0.25\mathbf{z}$, $\mathbf{z} \sim N(\mathbf{0}, \mathbf{I})$;
Variation 2: $\mathbf{x}_1, \mathbf{x}_2, \cdots, \mathbf{x}_{19} \sim N(\mathbf{0}, \mathbf{I})$, $\mathbf{x}_{20} = \mathbf{x}_{10} + 0.25\mathbf{z}$, $\mathbf{z} \sim N(\mathbf{0}, \mathbf{I})$;
Variation 3: $\mathbf{x}_1, \mathbf{x}_2, \cdots, \mathbf{x}_{19} \sim N(\mathbf{0}, \mathbf{I})$, $\mathbf{x}_{20} = \mathbf{x}_{15} + 0.25\mathbf{z}$, $\mathbf{z} \sim N(\mathbf{0}, \mathbf{I})$;
Variation 4: $\mathbf{x}_j = \mathbf{z} + \epsilon_j, j = 1, 2, \cdots, 20$, $\epsilon_j \sim N(\mathbf{0}, \mathbf{I})$, $\mathbf{z} \sim N(\mathbf{0}, \mathbf{I})$.

Table 1 lists the value of σ and the correlation structure used for the above 5 variations. Notice that variation 0 is the easiest problem since all the variables are independent. In variations 1–3, $\mathbf{x}_{20}$ is highly correlated with one of the three useful variables. Through these three variations, we can investigate the behavior of BSSW when the problem is highly collinear. In variation 4, high pairwise correlations are introduced among all the variables. For each variation, we repeated the simulation for 100 times. For a method $\mathcal{M}$, both the "soft" and "hard" metrics utilized in [14] were adopted to assess its performance. Interested readers can consult [14] for the detailed explanation and calculation of these two metrics. In current context, the "hard" metric assesses how well $\mathcal{M}$ performs to select significant variables whereas the "soft" metric evaluates its performance to rank the variables according to their importance.

Table 1. The σ value and correlation structure for the 5 variations.

Scenario	σ	Correlation stucture
Variation 0	1	$\rho(\mathbf{x}_j, \mathbf{x}_k) = 0$, $\forall j \neq k$
Variation 1	1	$\rho(\mathbf{x}_j, \mathbf{x}_k) \approx 0.97$ for $j = 5$ and $k = 20$
Variation 2	1	$\rho(\mathbf{x}_j, \mathbf{x}_k) \approx 0.97$ for $j = 10$ and $k = 20$
Variation 3	1	$\rho(\mathbf{x}_j, \mathbf{x}_k) \approx 0.97$ for $j = 15$ and $k = 20$
Variation 4	2	$\rho(\mathbf{x}_j, \mathbf{x}_k) \approx 0.50$, for all $j \neq k$

In this experiment, the ensemble size B was taken to be 25. With regard to PGA, another parameter N, that is, the number of generations for each SGA to evolve, need to be specified beforehand. In our experiment, the strategy proposed in [14] was used. In particular, we first ran SGA for 10 times using different initial populations and found that SGA reaches convergence by evolving about 20 generations on average. Hence, the parameter N in PGA was set to be 10. To make the comparison fair, the number of generations for SGA was taken to be 250 so that its total amount of computation and that of PGA is almost the same. The stepwise selection was realized with the "Stats" package of Matlab. Moreover, AIC and BIC criteria were both utilized to perform variable selection for all the studied methods. Because there is little difference between the results obtained with them, only those calculated with BIC are reported. Table 2 lists the hard and soft metric values of various methods for each variation.

The results in Table 2 leads to the following conclusions. First, SGA is hopeless to identify the correct set of variables no matter whether hard or soft metric is considered. Second, the ensemble approaches greatly enhance the performance of traditional selection methods. For instance, stepwise selection method performs badly to identify the right model. However, BSS and BSSW behaves quite well to select the truly important variables. The situation is similar when comparing SGA with PGA. In addition, the advantage of BSSW over BSS indicates that assigning

Table 2. The performance of various methods measured with *hard* and *soft* metrics for 5 variations.

Type of metric	Method	Variation 0	Variation 1	Variation 2	Variation 3	Variation 4
Hard	SGA	0.27	0.14	0.23	0.29	0.20
	PGA	0.86	0.06	0.66	0.91	0.37
	BSSW	0.93	0.37	0.68	0.82	0.28
	BSS	0.76	0.32	0.54	0.78	0.44
	Stepwise	0.00	0.00	0.00	0.00	0.00
Soft	SGA	0.59	0.35	0.63	0.67	0.60
	PGA	1.00	0.53	0.90	0.99	0.71
	BSSW	0.99	0.67	0.89	0.98	0.68
	BSS	0.98	0.57	0.84	0.96	0.65
	Stepwise	0.99	0.71	0.90	0.97	0.73

a mixed importance measure to each variable works better than only considering whether they occur in the final model. Thirdly, BSSW is observed to behave best for variations 0–2. For the other two variations, it is competitive with the best method. Here, it's noteworthy that variation 1 is the hardest case for all methods. In that situation, x_{20} is made to be highly correlated with x_5 which has the smallest nonzero coefficient relative to the noise σ in the true model. However, BSSW performs significantly better to detect the variables that actually affect the response y. Therefore, BSSW can be a highly competitive variable selection tool and it is relatively easy to use since it has only one parameter (namely, ensemble size B) for users to specify. In contrast, PGA has another parameter N need to be determined except for B.

It is worthwhile mentioning that another phenomenon emerges in Table 2. One may find that the performance of Stepwise is worst among the considered algorithms in terms of hard metric. However, it works satisfactorily in comparison with that of VSEs when evaluating each algorithm with soft metric. This shows that Stepwise can generally sort the variables in the order that is consistent with their true importance. However, it cannot choose all the variables which has actual effect on the response. Put in another way, stepwise selection often misses some actually influential variables or falsely includes some spurious variables even though it can accurately rank the variables in line with their relative importance to the response variable. As for BSSW, it can not only rank the variables in the correct order but also identify the right model. Finally, we would like to state that the conclusion of Stepwise drawn here is obtained just on the currently considered data sets. With regard to its performance on the other regression problems, it needs to be investigated further on a broad class of data sets.

4.2   Simulation 2

Here, we considered a widely used benchmark simulation [6,9,12,15,21]. There are $p = 8$ variables and each one is generated from the standard normal distribution. Furthermore, the pairwise correlation between two variables is $\rho(x_i, x_j) = 0.5^{|i-j|}$ for all $i \neq j$. The response y is generated by

$$y = 3x_1 + 1.5x_2 + 2x_5 + \sigma\epsilon, \quad \epsilon \sim N(\mathbf{0}, \mathbf{I}). \tag{6}$$

In this situation, only three variables (i.e., variables 1, 2 and 5) are truly important and the rest five ones are unimportant. This benchmark was first used in [12], but ever since it has been employed by many researchers to test the behavior of a variable selection technique.

First, we considered the case $n = 40$ and $\sigma = 3$. Then, we reduced σ to 1 and increased the sample size to 60. Under each situation, the experiment was repeated for 100 times for each method. As for the parametric setup, it was similar to that used in Simulation 1. The number of evolving generations for SGA was taken to be 250. For the procedures to construct a VSE, the ensemble size B was set to be 25. In PGA, each SGA was evolved for only 10 generations so that it has not reach convergence. In this way, the diversity among the ensemble members can be ensured.

Table 3. The performance of each method for the widely used benchmark simulation.

Method	Average number of zero coefficients		Average model size	Average PE ratio	Average AUC
	$x_j \in$ IM group $(j = 1, 2, 5)$	$x_j \in$ UIM group $(j = 3, 4, 6, 7, 8)$			
Oracle	0.00	5.00	3.00	1.00	1.00
$n = 40, \sigma = 3$					
SGA	2.06	3.42	2.52	2.3894	0.4987
PGA	0.90	4.91	2.19	1.3582	0.8400
BSSW	1.04	4.98	1.98	1.4132	0.8548
BSS	0.72	4.89	2.39	1.3123	0.8019
$n = 60, \sigma = 1$					
SGA	2.27	3.42	2.31	14.4233	0.4623
PGA	0.50	4.97	2.53	1.9885	0.8625
BSSW	0.00	5.00	3.00	1.0769	0.9993
BSS	0.00	4.87	3.13	1.0907	0.9157

In order to evaluate the performance of each method, we recorded the average number of zero coefficients respectively for the important group (IM group, $j = 1, 2, 5$) and unimportant variable group (UIM group, $j = 3, 4, 6, 7, 8$). Notice that the first term (number of incorrect zeros) actually characterizes the method's under-fitting effect while the second one (number of incorrect zeros) characterizes the method's capability in producing sparse models. We also reported the average model size, namely, the mean size of the selected model over 100 runs of experiments. The corresponding results for all the methods are summarized in Table 3. The method "Oracle" refers to fitting the model while pretending that we knew in advance the true model contains only the variables x_1, x_2 and x_5. In the meantime, we utilized the variables selected by each method to build a linear regression model. Then, the corresponding predictor error was computed using a test set whose size is identical to that of the training set. Here, the average ratio of the prediction error of each method to that of the true model is also displayed (see the penultimate column of Table 3). Furthermore, the AUC measure utilized in [2] was also taken into account. Here, it should be cautious to understand the results of the average model size. Although the close relationship between its value and the oracle value indicates the good performance of an algorithm, the variables having nonzero coefficients do not necessarily coincide with the underlying important ones.

From the results reported in Table 3, the following conclusions can be yielded. When the noisy level of the training data is low, BSSW performs best in terms of each evaluation measure. In particular, the average AUC value for BSSW is much higher than the other methods in each case. Among the compared methods, the performance of SGA is the worst since it more often omits some important variables or selects some additional unimportant variables. Although PGA and

Table 4. Variable selection frequencies of different methods for the widely used benchmark simulation.

Method	$x_j \in$ IM group $(j = 1, 2, 5)$			$x_j \in$ UIM group $(j = 3, 4, 6, 7, 8)$		
	Minimum	Median	Maximum	Minimum	Median	Maximum
Oracle	100	100	100	0	0	0
$n = 50, \sigma = 1$						
SGA	28	33	38	24	27	39
PGA	58	98	100	0	0	0
BSSW	100	100	100	0	0	0
BSS	100	100	100	1	2	4
$n = 50, \sigma = 3$						
SGA	26	32	32	25	30	41
PGA	40	82	96	0	0	1
BSSW	43	70	96	0	0	1
BSS	63	90	100	0	2	4
$n = 50, \sigma = 6$						
SGA	23	26	36	27	31	35
PGA	34	37	76	1	4	6
BSSW	30	43	76	1	4	8
BSS	43	45	80	3	5	9

BSSW perform comparably to exclude unimportant variables, PGA may falsely categorize some truly influential variables into uninfluential ones since the average number of zero coefficients for the variables x_1, x_2 and x_5 is large than zero. In comparison with BSS, BSSW works better to identify the variables which in fact have no impact on the response. When the sample size changes to be smaller and the training data contain more noise, the behavior of each method decreases to some degree. Under this situation, BSSW is observed to have the competitive performance with PGA while both of them work better than SGA and BSS. From the results of the average ratios of prediction error, it can be seen that the performance of PGA, BSSW and BSS is comparable, whereas SGA performs very badly.

As a second part of this experiment, we recorded the variable selection frequencies of different methods for several cases. The detailed results are reported in Table 4. Here, we list the minimum, median and maximum number of times out of 100 simulations among all important or unimportant variables are selected, respectively. To facilitate the understanding of these statistics, we will briefly describe how these frequencies are computed. It should be noticed that we repeated the experiment for 100 times for each combination of n and σ. In each replication, the frequency associated with a variable is increased by 1 if this variable is considered to be important by an algorithm. Evidently, the frequency for a variable which is more often selected will approximate 100. Because the experimental data are artificially generated, we know in advance that the indices of truly important and unimportant variables. For important variable group (i.e., $j = 1, 2, 5$), we can thus compute the minimal, median and maximum number of times that these variables are identified to be important. For unimportant variable group (i.e., $j = 3, 4, 6, 7, 8$), the similar calculation can be executed. Ideally, the frequencies for important variables should have high values close to 100 whereas the frequencies for unimportant ones should have low values close to 0.

From Table 4, we can draw some conclusions analogous to those previously obtained. Specifically, BSSW and PGA are comparative to accurately select the actually influential variable while the ability of PGA to exclude unimportant variables is a little weaker. SGA is indeed not a good technique to perform variable selection since the frequencies that it selects the important and unimportant variables are far away from those corresponding to Oracle. As for BSS, it seems that it performs comparatively with BSSW and PGA to identify the important variables. However, it may falsely consider some unimportant variables as important ones. Hence, this simulation study provides strong empirical evidence that the proposed method BSSW can be a highly competitive variable selection tool, especially when the noisy level of the training data is low.

5   Conclusions

In this paper, we developed a novel method called BSSW to construct a VSE to obtain higher selection accuracy. The main idea of BSSW is to draw some bootstrap samples and apply stepwise selection algorithm to each of them. Based on the order that one variable is selected into the final model as well as its contribution to model fitting, we assign a weight to every variable to reflect its relative importance to the response variable. Eventually, the average importance measure for each variable is obtained by averaging the results over bootstrapping trials and the variables are then ordered. If the goal is to detect significant variables for interpretation, a further step such as thresholding or some other technique can be performed. The conducted experiments demonstrate that BSSW performs better than some other variable selection techniques, especially in excluding unimportant variables. Additionally, it is easy to use because users only need to specify one parameter (namely, ensemble size B) in advance. Thus, BSSW can behave as an effective tool to implement variable ranking and selection in linear regression models. In the future, it is interesting to extend BSSW to generalized linear models and other more complex models.

Acknowledgements. This research was supported by the National Basic Research Program of China (973 Program, No. 2013CB329406), the National Natural Science Foundations of China (No. 11201367, 91230101), the Science Plan Foundation of the Education Bureau of Shaanxi Province of China (No. 14JK1672).

References

1. Bühlmann, P., Mandozzi, J.: High-dimensional variable screening and bias in subsequent inference, with an empirical comparison. Comput. Stat. **29**(3–4), 407–430 (2014)
2. Liu, C., Shi, T., Lee, Y.: Two tales of variable selection for high dimensional regression: screening and model building. Stat. Anal. Data Min. **7**(2), 140–159 (2014)
3. Fan, J.Q., Lv, J.C.: A selective overview of variable selection in high dimensional feature space. Stat. Sinica **20**(1), 101–148 (2010)
4. Wasserman, L., Roeder, K.: High-dimensional variable selection. Ann. Stat. **37**(5A), 2178–2201 (2009)
5. Shmueli, G.: To explain or to predict? Stat. Sci. **25**(3), 289–310 (2010)
6. Xin, L., Zhu, M.: Stochastic stepwise ensembles for variable selection. J. Comput. Graph. Stat. **21**(2), 275–294 (2012)
7. Breiman, L.: Heuristics of instability and stabilization in model selection. Ann. Stat. **24**(6), 2350–2383 (1996)
8. Miller, A.: Subset Selection in Regression (Second Edition). Chapman & Hall/CRC Press, New Work (2002)
9. Fan, J.Q., Li, R.Z.: Variable selection via nonconcave penalized likelihood and its oracle properties. J. Am. Stat. Assoc. **96**(456), 1348–1360 (2001)
10. Meinshausen, N., Bühlmann, P.: Stability selection (with discussion). J. Royal Stat. Soc. (Ser. B) **72**(4), 417–473 (2010)
11. Shah, R.D., Samworth, R.J.: Variable selection with error control: another look at stability selection. J. Royal Stat. Soc. (Ser. B) **75**(1), 55–80 (2013)
12. Tibshirani, R.: Regression shrinkage and selection via the lasso. J. Royal Stat. Soc. (Ser. B) **58**(1), 267–288 (1996)
13. Zou, H.: The adaptive lasso and its oracle properties. J. Am. Stat. Assoc. **101**(476), 1418–1429 (2006)
14. Zhu, M., Chipman, H.A.: Darwinian evolution in parallel universes: a parallel genetic algorithm for variable selection. Technometrics **48**(4), 491–502 (2006)
15. Wang, S.J., Nan, B., Rosset, S., Zhu, J.: Random lasso. Ann. Appl. Stat. **5**(1), 468–485 (2011)
16. Zhou, Z.H.: Ensemble Methods: Foundations and Algorithms. Taylor & Francis, Boca Raton (2012)
17. Breiman, L.: Bagging predictors. Mach. Learn. **24**(2), 123–140 (1996)
18. Freund, Y., Schapire, R.: A decision-theoretic generalization of on-line learning and an application to boosting. J. Comput. Sys. Sci. **55**(1), 119–139 (1997)
19. Friedman, J.H.: Greedy function approximation: a gradient boosting machine. Ann. Stat. **29**(5), 1189–1232 (2001)
20. Mendes-Moreira, J., Soares, C., Jorge, A.M., de Sousa, J.F.: Ensemble approaches for regression: a survey. ACM Comput. Surv. **45**(1), 40 (2012). Article 10
21. Zhang, C.X., Wang, G.W., Liu, J.M.: RandGA: injecting randomness into parallel genetic algorithm for variable selection. J. Appl. Stat. **42**(3), 630–647 (2015)
22. Zhu, M., Fan, G.Z.: Variable selection by ensembles for the Cox model. J. Stat. Comput. Simul. **81**(12), 1983–1992 (2011)

A Hierarchical Ensemble Method for DAG-Structured Taxonomies

Peter N. Robinson[1,4], Marco Frasca[2], Sebastian Köhler[1], Marco Notaro[3], Matteo Re[2], and Giorgio Valentini[2(✉)]

[1] Institut fur Medizinische Genetik und Humangenetik, Charité - Universitatsmedizin Berlin, Berlin, Germany
{peter.robinson,sebastian.koehler}@charite.de
[2] AnacletoLab - DI, Dipartimento di Informatica, Università degli Studi di Milano, Milan, Italy
{valentini,frasca,re}@di.unimi.it
[3] Dipartimento di Bioscienze, Università degli Studi di Milano, Milan, Italy
marco.notaro@studenti.unimi.it
[4] Institute for Bioinformatics, Department of Mathematics and Computer Science, Freie Universität Berlin, Berlin, Germany

Abstract. Structured taxonomies characterize several real world problems, ranging from text categorization, to video annotation and protein function prediction. In this context "flat" learning methods may introduce inconsistent predictions, while structured output-aware learning methods can improve the accuracy of the predictions by exploiting the hierarchical relationships between classes. We propose a novel hierarchical ensemble method able to provide theoretically guaranteed consistent predictions for any Directed Acyclic Graph (DAG)-structured taxonomy, and consequently also for any taxonomy structured according to a tree. Results with a complex real-world DAG-structured taxonomy involving about one thousand classes and twenty thousand of examples show that the proposed hierarchical ensemble approach significantly improves flat methods, especially in terms of precision/recall curves.

Keywords: Hierarchical ensemble classification methods · DAG-structured prediction · Multi-label classification

1 Introduction

Structured output classification consists in the prediction of multiple labels that are hierarchically correlated according to a pre-defined data structure, e.g. a tree or a directed acyclic graph (DAG). In this context "flat" classification methods, that predict labels independently of each other, can in principle be applied, but may introduce significant inconsistencies in the classification, due to the violation of the *true path rule* (also known as the *annotation propagation rule*) that governs the hierarchical relationships between classes [1,2]. According to this rule, a positive prediction for a class and a negative prediction for its parent

© Springer International Publishing Switzerland 2015
F. Schwenker et al. (Eds.): MCS 2015, LNCS 9132, pp. 15–26, 2015.
DOI: 10.1007/978-3-319-20248-8_2

classes are not allowed, since this violates the inclusion relationship between them. Therefore, a positive prediction for a class implies positive predictions for all of the ancestors of the class, and a negative prediction implies negative predictions for all of the class's descendants to avoid violating the true path rule. Moreover, flat methods do not take into account the hierarchical relationships between classes, thus loosing important a priori knowledge about the constraints of the hierarchical labeling.

To properly handle these problems, several structured output-aware learning methods have been proposed to exploit the a priori known relationships between labels. A first general approach is based on the kernelization of both the input and the output space, through the introduction of a joint kernel that computes the "compatibility" of a given input-output pair [4], or through other related techniques based on large margin methods for structured and interdependent output variables [3,5]. A recent work showed also that structured output methods can be enhanced by combining them through relatively simple ensemble techniques [6].

A second general approach is based on ensemble methods able to exploit the hierarchical relationships between classes [7]. More precisely, hierarchical ensemble methods, in their more general form, adopt a two-step learning strategy. In the first step each base learner separately or interacting with connected base learners learns a specific class. In most cases this yields a set of independent classification problems, where each base learning machine is trained to learn a specific class, independently of the other base learners. In the second step the predictions provided by the trained classifiers are combined by considering the hierarchical relationships between the base classifiers modeled according to the hierarchy of the classes.

Most of the proposed hierarchical ensemble methods focused on tree-structured taxonomies [7–10] and the ones specific for DAGs [1,11] showed that it is difficult to improve upon flat predictions.

We propose a novel ensemble learning strategy that exploits the DAG structure of the taxonomy through a double flow of information between the base learners associated to each class/node of the hierarchy: after separately learning each class with a specific classifier, predictions are first combined from bottom to top to enhance sensitivity, and successively from top to bottom to improve the precision of the predictions.

We provide theoretical guarantees that the proposed True Path Rule (*TPR-DAG*) hierarchical ensembles obey the true path rule in DAGs. Moreover we experimentally show that our approach can consistently improve flat predictions in a complex task involving human gene - phenotype associations, where classes are DAG-structured according to the Human Phenotype Ontology (HPO) [12].

2   True Path Rule (*TPR-DAG*) Hierarchical Ensembles for DAG Structured Taxonomies

TPR-DAG requires a first phase in which any class is learned by a dedicated base learner: in principle any base learner can be used to score each example.

After this learning phase, the second phase modifies these "flat" predictions to provide the *TPR* ensemble predictions. This second phase is divided into two steps:

1. *Bottom-up step.* For each example the DAG is visited from bottom to top to propagate "positive" predictions across the hierarchy. The aim of this step is to enhance the sensitivity of the predictions.
2. *Top-down step.* Starting from the root, and traversing the DAG toward the bottom, "negative predictions" are propagated toward the children. The aim of this step is to enhance the precision of the predictions.

This method builds on the previously proposed *TPR* ensemble method that can be safely applied only to tree-structured taxonomies [9,13]. The main difference with respect to the original tree-version consists in the fact that the per-level traversal is now performed through two completely distinct steps: a bottom-up per level visit of the graph followed by a top-down visit, while in the original tree-version the per-level traversal is performed in an "interleaved" fashion (that is the bottom-up and top-down traversal are alternated at each level [9]). Moreover the level of a class is defined in terms of its maximum distance from the root, since in a DAG we may have multiple paths from each node to the root. These two items (bottom-up and top-down separation and levels defined in terms of the maximum distance from the root) assure the true path rule consistency of the predictions, i.e. the requirement that the score of a parent or an ancestor node must be larger or equal than that of its children or descendants.

In the next subsections, after introducing some basic notations and definitions, we describe in detail the bottom-up and top-down steps of the *TPR-DAG* algorithm, as well its consistency properties.

2.1 Basic Notation and Definitions

Let $G =< V, E >$ denote a Directed Acyclic Graph (DAG) with vertices $V = \{1, 2, \ldots, |V|\}$ and edges $e = (i, j) \in E, i, j \in V$, where nodes $i \in V$ represent classes of the taxonomy and a direct edge $(i, j) \in E$ the hierarchical relationship between i and j: i is the parent class and j is the child class. The set of children of a node i is denoted $child(i)$, the set of its parents $par(i)$, the set of its ancestors $anc(i)$ and the set of its descendants $desc(i)$.

A "flat continuous" classifier $f_i : X \rightarrow [0, 1]$ associated with each node $i \in V$ provides scores $\hat{y}_i \in [0, 1]$ that can be interpreted as the likelihood or probability for a given example $x \in X$ of belonging to a given class i. The set of $|V|$ flat classifiers provides a multi-label score $\hat{\boldsymbol{y}} \in [0, 1]^{|V|}$:

$$\hat{\boldsymbol{y}} =< \hat{y}_1, \hat{y}_2, \ldots, \hat{y}_{|V|} > \tag{1}$$

We say that a multi-label scoring $\boldsymbol{y}$ is consistent if it obeys the *true path rule*:

$$\boldsymbol{y} \text{ is consistent} \iff \forall i \in V, j \in par(i) \Rightarrow y_j \geq y_i \tag{2}$$

2.2 Bottom-Up Step

The basic *TPR-DAG* adopts a per-level bottom-up traversal of the DAG, starting from the nodes most distant (in the sense of the maximum distance) from the root. More precisely, if $p(r, i)$ represents a path from the root node r and a node $i \in V$, $l(p(r, i))$ the length of a path p, $\mathcal{L} = \{0, 1, \ldots, \xi\}$ the set of observed levels, with ξ the maximum node level, then $\psi : V \longrightarrow \mathcal{L}$ is a level function which assigns each node $i \in V$ to its level $\psi(i)$:

$$\psi(i) = \max_p l\left(p(r, i)\right) \tag{3}$$

At each level the flat predictions $\hat{y}_i$ are changed to $\tilde{y}_i$ taking into account the "positive" predictions of its children:

$$\tilde{y}_i := \frac{1}{1 + |\phi_i|}(\hat{y}_i + \sum_{j \in \phi_i} \tilde{y}_j) \tag{4}$$

where ϕ_i are the "positive" children of i. The main goal of the bottom-up step consists in improving the sensitivity (recall) of the predictions. This is accomplished by allowing only the "positive" children (that is the nodes for which a relatively large score has been achieved) to transmit their scores to their parents. In this context a key issue is the selection of the positive children ϕ_i, and different strategies to select them can be applied:

1. *Threshold Free (TPR-TF) strategy.* A simple solution consists in choosing those children that can increment the score of the node i (that is positive nodes are those that achieve a higher score than that of their parent):

$$\phi_i := \{j \in child(i) | \tilde{y}_j > \hat{y}_i\} \tag{5}$$

2. *Thresholded (TPR-T) strategy.*
   In this case we set a threshold to select the positive children. We can a priori select a given threshold $\bar{t} \; \forall i \in V$, or we can select the threshold to maximize some performance metric estimated on the available data, e.g. the F-score or the AUC. The corresponding set of positives $\forall i \in V$ is:

$$\phi_i := \{j \in child(i) | \tilde{y}_j > \bar{t}\} \tag{6}$$

For instance $\bar{t}$ can be selected from a set of $t \in (0, 1)$ through cross-validation techniques.

Moreover we can also balance the weight $w \in [0, 1]$ between the prediction of the classifier associated with the node i and that of its "positive" children ϕ_i, through their convex combination. In this way, analogously to the "tree" version of the weighted *TPR* ensemble method [14] we can obtain the "weighted" version *TPR-W* of the *TPR-DAG* algorithm:

$$\tilde{y}_i := w\hat{y}_i + \frac{(1 - w)}{|\phi_i|} \sum_{j \in \phi_i} \tilde{y}_j \tag{7}$$

Independently of the variants of the basic *TPR-DAG* ensemble method, predictions are bottom-up propagated, thus moving positive predictions towards the parents and recursively towards the ancestors of each node.

2.3   Top-Down Step

The successive top-down step modifies the "bottom-up" scores computed in the previous bottom-up step (Sect. 2.2) by running in the opposite direction from the top to the bottom of the DAG. The main goal of this step consists in propagating "negative" predictions towards the children and recursively toward the descendants of each node, in order to provide consistent and more precise predictions. It adopts this simple rule by per-level visiting the nodes from top to bottom:

$$\bar{y}_i := \begin{cases} \tilde{y}_i & \text{if} \quad i \in root(G) \\ \min_{j \in par(i)} \bar{y}_j & \text{if} \quad \tilde{y}_i > \min_{j \in par(i)} \bar{y}_j \\ \tilde{y}_i & \text{otherwise} \end{cases} \tag{8}$$

The $\tilde{y}_i$ scores are those computed in the bottom-up step, while $\bar{y}_i$ are the final scores computer by the TPR ensemble.

The top-down step assures the hierarchical consistency of the predictions of the TPR, as stated by the following theorem:

Theorem 1. *Given a DAG $G =< V, E >$, a level function ψ that assigns to each node its maximum path length from the root, a set of predictions $\tilde{\boldsymbol{y}} =< \tilde{y}_1, \tilde{y}_2, \ldots, \tilde{y}_{|V|} >$ generated by the bottom-up step of the* TPR *algorithm for each class associated with its corresponding node $i \in \{1, \ldots, |V|\}$, the top-down step of the* TPR *algorithm assures that for the set of ensemble predictions $\bar{\boldsymbol{y}} =< \bar{y}_1, \bar{y}_2, \ldots, \bar{y}_{|V|} >$ the following property holds:*

$$\forall i \in V, \ j \in par(i) \Rightarrow \bar{y}_j \geq \bar{y}_i$$

The proof can be obtained by applying (8) to each node according to a per-level visit of the DAG, where levels are defined in terms of the maximum path length from the root (3), and by observing that each node is visited only once by the top-down step of the algorithm (details are omitted for lack of space).

From Theorem 1 it is easy to prove that the consistency of the predictions holds for all the ancestors of a given node $i \in V$:

Corollary 1. *Given a DAG $G =< V, E >$, the level function ψ, a set of flat predictions $\hat{\boldsymbol{y}} =< \hat{y}_1, \hat{y}_2, \ldots, \hat{y}_{|V|} >$ for each class associated with each node $i \in \{1, \ldots, |V|\}$, the* TPR *algorithm assures that for the set of ensemble predictions $\bar{\boldsymbol{y}} =< \bar{y}_1, \bar{y}_2, \ldots, \bar{y}_{|V|} >$ the following property holds:*

$$\forall i \in V, \ j \in anc(i) \Rightarrow \bar{y}_j \geq \bar{y}_i$$

The proof can be easily obtained from Theorem 1 by "reductio ad absurdum".

The function ψ that computes the maximum distance of each node from the root (Eq. 3) can be implemented through a straightforward variant of the classical Bellman-Ford algorithm [15]: by recalling that it finds the shortest paths from a source node to all the other nodes of a weighted digraph, it is sufficient to invert the sign of each edge weight to obtain the maximum distance (longest path) from the root. The complexity of the Bellman-Ford algorithm is cubic with respect the number of vertices, but recalling that the function ψ must be computed only once for a given hierarchical task, this complexity could be acceptable for most low and medium-sized DAGs. For big DAGs a variant of the classical topological sort algorithm for graphs can be applied instead: by exploiting the topological ordering of the nodes, the maximum distance from the root can be easily computed with time complexity $\mathcal{O}(|V| + |E|)$, that is in quadratic time for dense graph and in linear time for sparse DAGs with respect to the number of vertices.

2.4 The Overall *TPR-DAG* Algorithm

Figure 1 shows the high-level pseudo-code of the *TPR-DAG* algorithm. The first four rows compute the maximum distance of each node from the root, using the Bellman-Ford algorithm. Note that the with a certain abuse of notation $E' := \{e'|e \in E, e' = -e\}$ indicates a new set E' of edges having weights with opposite sign with respect to the original set of edges E. The block B (rows 5–12) performs a bottom-up visit of the graph and updates the predictions $\tilde{y}_i$ of the *TPR* ensemble according to Eq. 4 and one of the positive selection strategies described in Sect. 2.2. Note that this step propagates the "positive" predictions from bottom to top of the DAG, but does not assure their true path rule consistency. This is accomplished by the third block (rows 13–24) that simply executes a hierarchical top-down step, according to the procedures described in Sect. 2.3.

It is easy to verify that complexity of the *TPR* algorithm is $\mathcal{O}(|V|)$ for both the B and C blocks when the DAG is sparse, while the complexity of block A depends on the selected algorithm: by choosing the variant of the Bellman-Ford algorithm the complexity is $\mathcal{O}(|V|^3)$, while by applying the variant of the topological sort algorithm the complexity is $\mathcal{O}(|V|+|E|)$. Note that block A must be executed only once for all the examples, while blocks B and C must be iterated for each example whose DAG-structured multi-label should be predicted.

3 Experimental Set-Up

We applied the proposed hierarchical ensemble methods to the prediction of Human Phenotype Ontology (HPO) terms associated with Mendelian disease genes [16]. The HPO aims at providing a standardized categorization of the abnormalities associated with human diseases and the semantic relationships between them. More precisely, HPO classes (terms) describe human phenotypic abnormalities and are structured according to a DAG, where children terms can be interpreted as subclasses of their parents. The experiments presented in this manuscript are based on the September 2013 HPO release (10, 099 terms and 13, 382 between-term relationships). We downloaded from the same HPO release all the available annotations (gene-term associations), resulting in set of 2759 genes having at least 1 annotation. In our experiments we included a set of 20257 human genes, and hence more than 17000 genes had no HPO annotations.

```
Input:
- G =< V, E >
- V = {1, 2, ..., |V|}, 1 is the root node
- ŷ =< ŷ₁, ŷ₂, ..., ŷ_{|V|} >,   ŷ_i ∈ [0, 1]
begin algorithm
01:    A. Compute ∀i ∈ V the max distance from root(G):
02:        E' := {e'|e ∈ E, e' = −e}
03:        G' :=< V, E' >
04:        dist := Bellman.Ford(G', root(G'))
05:    B. Per-level bottom-up visit of G:
06:        for each d from max(dist) to 0 do
07:            N_d := {i|dist(i) = d}
08:            for each i ∈ N_d do
09:                Select φ_i according to a positive selection strategy
10:                ỹ_i := (1/(1+|φ_i|))(ŷ_i + Σ_{j∈φ_i} ỹ_j)
11:            end for
12:        end for
13:    C. Per-level top-down visit of G:
14:        ȳ₁ := ỹ₁
15:        for each d from 1 to max(dist) do
16:            N_d := {i|dist(i) = d}
17:            for each i ∈ N_d do
18:                x := min_{j∈par(i)} ȳ_j
19:                if (x < ỹ_i)
20:                    ȳ_i := x
21:                else
22:                    ȳ_i := ỹ_i
23:            end for
24:        end for
end algorithm
Output:
- ȳ =< ȳ₁, ȳ₂, ..., ȳ_{|V|} >
```

Fig. 1. Hierarchical true path rule algorithm for DAGs (TPR-DAG)

After pruning HPO terms having less than 50 annotations we obtained a final set of 911 HPO terms and $1,095$ between-term relationships that were used in our experiments.

A collection of feature vectors containing functional and biomolecular signatures describing the products of $20,257$ human genes was constructed starting from different publicly available biological databases (Table 1). Then the binary feature vectors were used to construct $n = 8$ gene networks (one for each data source listed in Table 1) by computing the Jaccard similarity between each possible pair of feature vectors associated to the genes.

Table 1. Data sources used in the experiments

Database	Content	Web site
InterPro	functional family, domains, functional sites	www.ebi.ac.uk/interpro
Pfam	functional family, domains	pfam.xfam.org
PRINTS	protein fingerprints, conserved motifs	www.bioinf.manchester.ac.uk
PROSITE	domains, families, functional sites	prosite.expasy.org
SMART	modular architectures	smart.embl-heidelberg.de
SUPFAM	structural and functional annotation	supfam.cs.bris.ac.uk
Gene Ontology	biological processes, cellular components and molecular functions	geneontology.org
OMIM	genetic diseases	www.omim.org
FI net (Wu et al.)	integrated network with expert-curated and non-curated sources of information	
HumanNet (Lee et al.)	integrated network with multi-species data	

We then combined the n gene networks by simply averaging the edge weights w_{ij}^d of each network $d \in \{1, n\}$ [17]:

$$\bar{w}_{ij} = \frac{1}{n} \sum_{d=1}^{n} w_{ij}^d \tag{9}$$

In order to construct a more informative gene network we performed the integration by adding two more functional gene networks (FI and HumanNet) taken from the literature [18,19], thus obtaining a final integration of 10 biomolecular networks (Table 1).

To process and provide flat scores for the considered 911 HPO terms using the above networked data we applied two semi-supervised methods: (a) the classical semi-supervised label propagation method (*LP*) based on Gaussian Fields and Harmonic Functions [20]; (b) the kernelized score functions (*RANKS*) semi-supervised network-based learning method recently successfully applied to both gene disease prioritization [17], and drug repositioning [21]. *RANKS* implements both local and global learning strategies by embedding in a "local" score function a graph kernel that takes into account the "global" topology of the network. In our experiments we applied *RANKS* with the average score function and the *1-step random walk kernel* [22].

4   Results

We compared the generalization performance of *Flat* and *TPR* ensemble methods by using 5-fold cross-validation techniques, and considering separately the two different base learners (*RANKS* and *LP*, Sect. 3). We also compared the results of *TPR* ensemble methods with three heuristic hierarchical ensemble methods (i.e. *And*, *Or* and *Max*), originally proposed for the hierarchical prediction of Gene Ontology terms [1]. It is worth noting that in the same work [1] *isotonic regression*-based hierarchical methods achieved better results than the heuristic ensemble algorithms used in our experiments, but we did non use them due to their computational complexity, considering the relatively large size of the taxonomy and of the input data considered here.

4.1 Experimental Results Using Kernelized Score Functions (*RANKS*) as Base Learner

By looking at the single 911 HPO terms (classes), in terms of AUC the *TPR-TF* ensemble achieves better results than *Flat* for 830 terms and worse results for 81 HPO terms. Table 2 shows that the average AUC across classes is only slightly larger for *TPR-TF* ensembles with respect to *Flat*, but the difference is statistically significant according to the Wilcoxon rank sum test. Also with respect to three heuristic hierarchical ensemble methods (*And*, *Or*, *Max*) *TPR-TF* achieve equal or significantly better results. More precisely the difference is statistically significant with respect to *Or* and *Max*, while no significant difference is registered with the *And* method.

Better results are obtained by the *TPR-TF* method in terms of the precision at fixed recall rates. Indeed the difference is statistically significant with respect to *Flat* and the three heuristic hierarchical ensemble methods, both at 10, 20 and 40 % recall (Table 2). These results are confirmed also by the precision-recall curves (Fig. 2): the *TPR-TF* solid line marked with circles is consistently above all the other curves, showing that *TPR-TF* achieves on the average better results than all the other methods compared.

Table 2. Average AUC, and precision at 10, 20 and 40 % recall (P10R, P20R and P40R), using kernelized score functions as base learner. Flat stands for flat ensemble method, TPR-TF for True Path Rule Threshold-Free, Max for Hierarchical Maximum, And for Hierarchical And and Or for Hierarchical Or ensemble methods. Methods that are significantly better than all the others according to the Wilcoxon rank sum test ($\alpha = 10^{-5}$) are highlighted in bold.

	Flat	TPR-TF	Max	And	Or
AUC	0.8213	**0.8269**	0.8241	**0.8274**	0.8241
P10R	0.2969	**0.3427**	0.2908	0.2815	0.2994
P20R	0.2043	**0.2333**	0.2025	0.1903	0.2081
P40R	0.1054	**0.1225**	0.1071	0.0993	0.1095

On the contrary, by comparing the different variants of the proposed *TPR* hierarchical ensemble methods, no statistically significant differences between them were identified (data not shown).

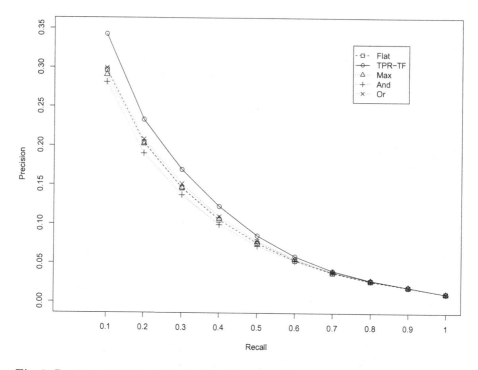

Fig. 2. Precision at different levels of recall, averaged across HPO terms (base learner: kernelized score functions)

4.2 Experimental Results Using Label Propagation (*LP*) as Base Learner

We repeated the same experiments using this time the label propagation (*LP*) method to implement the flat base learners. Also with this base learner we achieved significantly better results with *TPR* ensemble methods with respect to the *Flat* approach, both in terms of the average AUC and average precision at fixed recall rates. Especially considering precision at fixed recall rates the *TPR* ensemble achieved significantly better results than those obtained with *Flat* and the three heuristic hierarchical ensemble methods, according to the Wilcoxon rank sum test (Table 3).

It is worth noting that the absolute average AUC and precision values obtained with the *LP* base learner (Table 3) are significantly lower than those achieved with *RANKS* (Table 2), showing that *TPR* results, as well as those obtained by the other heuristic ensemble methods depend on the choice of the flat base learner. Nevertheless, *TPR* ensemble methods with *LP* base learners are able to achieve a relative precision improvement with respect to the *Flat* approach in the range between 15 and 40%, at least for recall rates between 0.1 and 0.4 (Table 3). Note that this is not the case for the three heuristic hierarchical ensemble methods (*And, Or, Max*), confirming previous results obtained in the context of gene function prediction problems [1].

Table 3. Average AUC, and precision at 10, 20 and 40 % recall (P10R, P20R and P40R), using label propagation as base learner. TPR-T stands for True Path Rule ensembles with Threshold. Methods that are significantly better than all the others according to the Wilcoxon rank sum test ($\alpha = 10^{-5}$) are highlighted in bold.

	Flat	TPR-T	Max	And	Or
AUC	0.7883	**0.7967**	0.7869	**0.7974**	0.7923
P10R	0.0673	**0.0936**	0.0653	0.0704	0.0730
P20R	0.0568	**0.0709**	0.0549	0.0564	0.0606
P40R	0.0439	**0.0503**	0.0426	0.0444	0.0462

5   Conclusions

Several real-world problems ranging from text classification to protein function prediction are characterized by hierarchical multi-label classification tasks. In this context flat methods may provide inconsistent predictions and more in general are not able to exploit the hierarchical constraints between classes.

We theoretically guarantee that *TPR-DAG* ensembles provide predictions that obey the true path rule in DAG-structured taxonomies, and we show in a large experiment involving the DAG-structured Human Phenotype Ontology that our proposed hierarchical ensembles consistently outperform flat methods, independently of the base learner used.

We outline that the proposed hierarchical method is independent of the base learner used, even if learners providing scores or probabilities of belonging to a given class are better suited for the *TPR-DAG* ensembles. From this standpoint *TPR* methods can be conceived as a flexible tool that can be applied to any off-the-shelf flat method to improve its predictions for DAG-structured taxonomies. *TPR-DAG* can be also applied also to tree-structured taxonomies, since obviously trees are DAGs.

This reseach could be extended by exploring other base learners, including also supervised learners, and by comparing *TPR* with other hierarchical methods, including also structured output kernel methods. To test the effectiveness of *TPR-DAG* ensembles in different application domains, the hierarchical classification of web documents and the protein function prediction problem could be two significant real-world test-beds for future experiments.

References

1. Obozinski, G. et al.: Consistent probabilistic output for protein function prediction. Genome Biol. 9(S6) (2008)
2. Robinson, P.N., Bauer, S.: Introduction to Bio-Ontologies. CRC Press, Boca Raton (2011)
3. Bakir, G., et al.: Predicting Structured Data. MIT Press, Cambridge (2007)
4. Lampert, C., Blaschko, M.: Structured prediction by joint kernel support estimation. Mach. Learn. **77**, 249–269 (2009)

5. Tsochantaridis, I., et al.: Large margin methods for structured and interdependent output variables. JMLR **6**, 1453–1484 (2005)

6. Cortes, C., Kuznetsov, V., Mohri, M.: Ensemble methods for structured prediction. In: Proceedings of the 31st ICML, Beijing, China (2014)

7. Silla, C., Freitas, A.: A survey of hierarchical classification across different application domains. Data Min. Knowl. Disc. **22**(1–2), 31–72 (2011)

8. Wang, H., She, X., Pan, W.: Large margin hierarchical classification with mutually exclusive class membership. JMLR **12**, 2649–2676 (2011)

9. Valentini, G.: True path rule hierarchical ensembles for genome-wide gene function prediction. IEEE ACM Trans. Comp. Biol. Bioinf. **8**(3), 832–847 (2011)

10. Cesa-Bianchi, N., Re, M., Valentini, G.: Synergy of multi-label hierarchical ensembles, data fusion, and cost-sensitive methods for gene functional inference. Mach. Learn. **88**(1), 209–241 (2012)

11. Schietgat, L., et al.: Predicting gene function using hierarchical multi-label decision tree ensembles. BMC Bioinformatics **11**(2), 1–14 (2010)

12. Robinson, P., et al.: The human phenotype ontology: a tool for annotating and analyzing human hereditary disease. Am. J. Hum. Genet. **83**, 610–615 (2008)

13. Valentini, G.: True path rule hierarchical ensembles. In: Benediktsson, J.A., Kittler, J., Roli, F. (eds.) MCS 2009. LNCS, vol. 5519, pp. 232–241. Springer, Heidelberg (2009)

14. Re, M., Valentini, G.: An experimental comparison of hierarchical bayes and true path rule ensembles for protein function prediction. In: El Gayar, N., Kittler, J., Roli, F. (eds.) MCS 2010. LNCS, vol. 5997, pp. 294–303. Springer, Heidelberg (2010)

15. Cormen, T., Leiserson, C., Rivest, R.: Introduction to Algorithms. MIT Press, Boston (2009)

16. Kohler, S., et al.: The human phenotype ontology project: linking molecular biology and disease through phenotype data. Nucleic Acids Res. **42**(D1), D966–D974 (2014)

17. Valentini, G., et al.: An extensive analysis of disease-gene associations using network integration and fast kernel-based gene prioritization methods. Artif. Intell. Med. **61**(2), 63–78 (2014)

18. Wu, G., Feng, X., Stein, L.: A human functional protein interaction network and its application to cancer data analysis. Genome Biol. **11**, R53 (2010)

19. Lee, I., et al.: Prioritizing candidate disease genes by network-based boosting of genome-wide association data. Genome Res. **21**(7), 1109–1121 (2011)

20. Zhu, X., et al.: Semi-supervised learning with gaussian fields and harmonic functions. In: Proceedings of the 20th ICML, Washintgton, DC, USA (2003)

21. Re, M., Valentini, G.: Network-based drug ranking and repositioning with respect to drugbank therapeutic categories. IEEE ACM Trans. Comp. Biol. Bioinf. **10**(6), 1359–1371 (2013)

22. Re, M., Valentini, G.: Cancer module genes ranking using kernelized score functions. BMC Bioinformatics **13**(Suppl 14/S3), 1–16 (2012)

Diversity Measures and Margin Criteria in Multi-class Majority Vote Ensemble

Ayako Mikami[(✉)], Mineichi Kudo, and Atsuyoshi Nakamura

Graduate School of Information Science and Technology,
Hokkaido University, Sapporo, Japan
{amikami,mine,atsu}@main.ist.hokudai.ac.jp

Abstract. Ensemble learning is a strong tool to strengthen weak classifiers. A large amount of diversity among those weak classifiers is a key to accelerate the effectiveness. Therefore, many diversity measures on a given training sample set have been proposed so far. However, they are almost all based on the oracle output that is one if the class predicted by the classifier is correct, zero otherwise. We point out such an oracle output scheme is not appropriate for the problems of more than two classes, and extend one of the most popular diversity measures, disagreement measure, to multi-class cases. On the other hand, the concept of margin has been recognized as an analytic tool to measure the generalization performance of a given classifier. Therefore, we analyze when some criteria for maximizing margins of an ensemble classifier over training samples are maximized under the assumption that the average accuracy of the base classifiers is constant. We also reveal the relationship between those criteria and the extended disagreement measure. As a result, it turns out that diversity is necessary not only over samples but also over predicted classes, if we want to extract the highest potential of ensemble.

1 Introduction

In the classifier design, estimating the generalization error of a classifier from training data is important. We consider in this paper the generalization error in ensemble classifiers that are made by combining many base classifiers. The beauty of ensemble by majority voting is that we can decrease the generalization error as we like by adding independent classifiers of error less than $1/2$. However, it is difficult to obtain independent classifiers in practice. In particular, there must be a correlation between classifiers trained by the same training data. Therefore, increasing the number of base classifiers does not always guarantee a better performance of the ensemble classifier. As an alternative of independence, many studies have been trying to increase the diversity of base classifiers, where the degree is measured by how differently these classifiers output different answers on training data. However, these diversity measures have been studied in the oracle output scheme [3] where the output of each base classifier is "correct" (1) or "wrong" (0). Unfortunately, as will be shown, such a scheme is not appropriate in multi-class (the number of classes is three or more)

© Springer International Publishing Switzerland 2015
F. Schwenker et al. (Eds.): MCS 2015, LNCS 9132, pp. 27–37, 2015.
DOI: 10.1007/978-3-319-20248-8_3

problems. Therefore, we reconsider how to measure the diversity in multi-class problems and investigate the relationship between a diversity measure and the generalization error through "margin" defined also in multi-class setting.

2   Definition

Let a training sample set be $\{(\boldsymbol{x}_1, y_1), \ldots, (\boldsymbol{x}_N, y_N)\}$, where x_i is a sample point in $\mathcal{R}^q$ and y_i is the class label taking one value in $\mathcal{K} = \{1, 2, \ldots, K\}$. Given L base classifiers $h_1, \ldots h_L$, let the output of jth classifier on sample x_i be $c_{ji} = h_j(x_i) \in \mathcal{K}$. That is, if $y_i = c_{ji}$ then x_i is correctly classified by h_j.

In addition, let the ratio of the number of classifiers voting to class k on sample x_i be q_{ik} as

$$\boldsymbol{q}_i = (q_{ik}), \ q_{ik} \triangleq \frac{1}{L} \sum_{j=1}^{L} \mathbb{I}(c_{ji} = k),$$

where $\mathbb{I}(\cdot)$ is the indicator function that takes one if the statement in the parenthesis is true and zero otherwise. In histogram $\boldsymbol{q}_i$, by t_i let us denote the class other than y_i taking the largest value, as

$$t_i = \max_{k \neq y_i} q_{ik}. \tag{1}$$

Then, as will be defined, the (training) margin m_i is defined as $m_i = q_{iy_i} - q_{it_i}$.

Last, the ensemble hypothesis h is produced by majority voting of the L classifiers and outputs y_i (if $q_{iy_i} \geq q_{it_i}$) or t_i (if $q_{iy_i} < q_{it_i}$). Thus, if $q_{iy_i} > q_{it_i}$, equivalently, $m_i > 0$, the ensemble classifier h correctly classifies x_i because of $y_i = h(x_i)$.

Each base classifier h_j has the (training) accuracy

$$p_j = \frac{1}{N} \sum_{i=1}^{N} \mathbb{I}(h_j(x_i) = y_i),$$

and the ensemble h has the average accuracy $p = \frac{1}{L} \sum p_j$. It is easy to confirm that p is also written as

$$p = \frac{1}{N} \sum_{i=1}^{N} q_{iy_i} \left(= \frac{1}{L} \sum_{j=1}^{L} p_j \right). \tag{2}$$

On the other hand, h has its own (training) accuracy p_h defined as

$$p_h = \frac{1}{N} \sum_{i=1}^{N} \mathbb{I}(q_{iy_i} > q_{it_i}).$$

In two-class cases, since the relation $q_{iy_i} > q_{it_i}$ means $q_{iy_i} > 1/2$, we have

$$p_h = \frac{1}{N} \sum_{i=1}^{N} \mathbb{I}(q_{iy_i} > \frac{1}{2}). \tag{3}$$

In the oracle output scheme, we have the same formula (3) by replacing q_{iy_i} with the ratio of the number of correct outputs to the number L of the total outputs. As a rough explanation for the secret of ensemble, comparing (2) and (3), we can say, if $p > 1/2$, implying almost all $q_{iy_i} > 1/2$, then p_h is close to one in (3).

3 Analysis of Advantage of Majority Votes

Here, we make a simple analysis for when the combined classifier of L base classifiers works well. We consider a sample $(\boldsymbol{x}, y)$.

Let $\mathbf{r} = (r_1, r_2, \ldots, r_K)$ be the probabilities of a multinomial distribution related to x and $\mathbf{q} = (q_1, q_2, \ldots, q_K)$ be the estimate from L independent samples, supposing the outputs of L independent base classifiers, generated according to the multinominal distribution. Then, it is obvious that $\mathbf{q}$ converges to $\mathbf{r}$ in probability by the law of large numbers, as L increases.

Now let us assume that the correct class is $y = 1$ without loss of generality. Then, the decision on the basis of this estimate $\mathbf{q}$ is correct if $q_1 = \max(q_1, q_2, \ldots, q_K)$. Our question is the condition, especially on r_1, under which this decision is also *really correct*, for a sufficiently large number of L.

In $K = 2$, it is clear that the answer is when $r_1 = \max(r_1, r_2 = r_1 - 1)$, thus, the necessary and sufficient condition is $r_1 > 1/2$. However, the necessary condition becomes a little complex for $K > 2$. In principle, it is simple because the necessary and sufficient condition is $r_1 = \max(r_1, r_2, \ldots, r_K)$. As we can see in Fig. 1 for $K = 3$, the region satisfying this condition occupies one-third of the possible nonnegativite region (the shaded area). If $\mathbf{r}$ falls in the area, we can expect the convergence to the correct decision with increasing base classifiers in the sense that the probability $P(q_1 \neq \max(q_1, q_2, q_3))$ approaches to zero. Note that if $r_1 > 1/2$, then any $\mathbf{r}$ falls in this area regardless of the values of the other probabilities. However, the actual necessary condition on r_1 is $r_1 > 1/K$. Then, the next question is the probability of $\mathbf{r}$ falling in this area in the case of $1/K < r_1 < 1/2$ (see line L in Fig. 1). The answer is illustrated in Fig. 2, assuming the uniform distribution of $(r_2, r_3, \ldots, r_K)$ conditioned $r_2 + r_3 + \cdots r_K = 1 - r_1$. When $K = 3$, it is easy to show that the probability is given by

$$P(r_1 = \max(r_1, r_2, r_3)) = \frac{3r_1 - 1}{1 - r_1} \quad (1/3 < r_1 < 1/2).$$

It is noted that this probability becomes one for $r_1 = 1/2$ and zero for $r_1 = 1/3$. In another deterministic viewpoint, we can say that the range of r_1 as the necessary condition becomes larger as the remaining r_2 and r_3 approaches to a same value: for example, $r_1 > 1/3$ becomes the necessary and sufficient condition if $r_2 = r_3 = (1 - r_1)/2$.

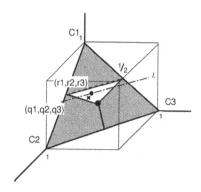

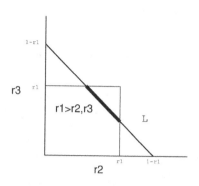

Fig. 1. Necessary condition for convergence to the Bayes classifier in case of multi-class cases $(K = 3)$

Fig. 2. Probability that r_1 dominates the others

As a result, this simple analysis says that the possibility of convergence to the correct decision of a sample is strengthened by a higher value of $r_1 > 1/K$ and by a higher level of the closeness of the remaining probabilities.

4   Related Works

Diversity of base classifiers is a practical measure of the effectiveness of ensemble. For evaluating the diversity of the base classifiers, several measures have been proposed so far [2,3]. In addition, some comparative studies [3] report that many measures behave similarly in many experiments. The problem is that almost all diversity measures are calculated in the oracle output scheme, that is, 0–1 outputs even for multi-class cases.

Recently, one of promising tools to measure the generalization performance of a classifier is the margin. In addition, the margin of ensemble classifiers have been studied [1,4]. They derived an upper bound of the generalization error in terms of the margins.

As a meeting point of the two streams of studies, a study has been made on the relationship between the margin and the measures of diversity in the context of ensemble learning [5]. Unfortunately, there was not a sufficient amount of correlation between them, but such a trial is still interesting. In this study, we follow this study and extend the discussion made in the oracle output setting to that in multi-class setting.

5   The Margin Analysis

5.1   Margin

As stated already, it is difficult to estimate the generalization error of a classifier accurately from a finite set of training data. Nevertheless, some theoretical stud-

ies have revealed the connection between the training error and the generalization error, mostly by showing an upper bound of the generalizing error in terms
of the training error with some complexity measure (e.g. Vapnik-Chervonenkis
dimension [4]) of the family to which the classifier belongs.

Such an upper bound is also shown in terms of *margin* on training samples [1,4].
For example, Schapire *et al.* [4] showed that

Theorem 1 ([4], Theorem 6). *Let $\mathcal{D}$ be a distribution over $X \times Y$, and let S be
a sample of N examples chosen independently at random according to $\mathcal{D}$. Assume
that the base-classifier space $\mathcal{H}$ is finite, and let $\delta > 0$. Then with probability at
least $1 - \delta$ over the random choice of the training set S, every function $f \in \mathcal{C}$
satisfies the following bound for all $\theta > 0$:*

$$P_{\mathcal{D}}[margin(f, x, y) \leq 0]$$
$$\leq P_S[margin(f, x, y) \leq \theta]$$
$$+ \mathcal{O}\left(\frac{1}{\sqrt{N}}\left(\frac{log(NK)log|\mathcal{H}|}{\theta^2} + log(1/\delta)\right)^{1/2}\right), \quad (4)$$

where $margin(f, x, y)$ is given by $f(x, y) - \max_{y' \neq y} f(x, y')$ (see [4] for the detail).

Here $P_{\mathcal{D}}$ and P_S are the probabilities over distribution $\mathcal{D}$ and the empirical sample distribution, respectively. In addition $\mathcal{C}$ is the convex hull of base classifiers.
This theorem says that the larger the margins are in the training data, the less
the generalized error is.

According to Breiman [1] and notation (1), let us define the margin m_i^1 of
sample x_i with class label y_i by

$$m_i = q_{iy_i} - q_{it_i}.$$

That is, the margin is the difference between the conditional probability of
the correct class and that of most likely but wrongly predicted class, given x_i.
If $m_i > 0$, then the prediction is correct.

In $K = 2$, that is, in two-class cases, $m_i = q_{iy_i} - q_{it_i} = 2q_{iy_i} - 1$. Thus,
from (2),

$$p = \frac{1}{N}\sum_{i=1}^{N} q_{iy_i} = \frac{1}{N}\sum_{i=1}^{N}\frac{m_i + 1}{2} = \frac{1}{2N}\sum_{i=1}^{N} m_i + \frac{1}{2}.$$

Hence, we have

$$\sum_{i=1}^{N} m_i = (2p - 1)N.$$

This means that the sum of margins is constant under a constant average
accuracy assumption of p and becomes positive only when $p > 1/2$. However,
this implication does not hold for $K > 2$, that is, constant p does not mean
constant sum of margins.

---

[1] The margin in (4) is connected to this m_i by $margin(f, x_i, y_i) = q_{iy_i} - q_{it_i} = m_i$.

5.2   Diversity and Margin

Some studies have shown that increasing of margins guarantees decreasing of the generalization error such as (4). A study [5], furthermore, analyzed the relationship between several diversity measures and the minimum margin $\underline{m} = \min_i m_i$, expecting that increasing $\underline{m}$ derives decreasing the generalization error. If we maximize the minimum margin $\underline{m}$ over training samples under the condition of $p = \text{const}$ and $K = 2$ in the oracle output scheme, that is, if we solve the following problem

$$\max_{\{m_i\}} \min_j \ m_j, \quad \text{subject to} \ \sum_{i=1}^{N} m_i = \text{const},$$

then the solution is given by $m_i = 2p - 1, \forall i \in \{1, 2, \ldots, N\}$. In the other words, the minimum margin is maximized if and only if every base classifier h_j classifies the training samples evenly. Note that this solution is obtained in the oracle output scheme or two-class cases only.

5.3   Extension to Multi-class Cases

Let us extend this discussion to the problems of three or more classes. The natural assumption is still a constant average accuracy p because p is the average of individual performance p_i's. In the multi-class setting, this assumption does not directly mean a constant sum of margins anymore, because $\sum_{i=1}^{N} m_i = \sum_{i=1}^{N} (q_{iy_i} - q_{it_i}) = \sum q_{iy_i} - \sum q_{it_i} = Np - \sum q_{it_i}$. That is, margin m_i is related to not only q_{iy_i} but also q_{it_i}. Note that, even so, the benefit of larger margins still holds.

Under $p = \text{const}$, or equivalently, $\sum q_{iy_i} = \text{const}$ (from (2)), we examine the best strategy to assign $\{q_{ik}\}$, or component-wisely $\{c_{ji}\}$, for a fixed N and L.

We consider three criteria of maximizing margins on the training data:

(1) $\underline{m} = \min_i m_i \rightarrow \max$,

(2) $\sum_{i=1}^{N} m_i \rightarrow \max$, and

(3) $\sum_{i=1}^{N} \mathbb{I}(m_i \geq \theta) \rightarrow \max$, for some $\theta > 0$.

First, we note that margin m_i is independent from other $m_{i'}$ ($i' \neq i$), $c_{ji'}$ ($i' \neq i$) and $q_{i'k}$ ($i' \neq i$). Therefore, it suffices to consider only c_{ji}'s to increase the value of m_i for fixed i.

It is important to notice that, for fixed q_{iy_i}, we can maximize the value of $m_i = q_{iy_i} - q_{it_i}$ by minimizing the (first or second largest) value of h_{it_i}, regardless of the value of q_{iy_i}. This is achieved by making all the values of $h_{ik}(k \neq y_i)$ be equal. Thus, in every criteria (1)–(3), the best local strategy is to have

$$q_{ik} = \frac{1 - q_{iy_i}}{K - 1}, \quad k = 1, 2, \ldots, K, \ k \neq y_i. \tag{5}$$

This also means $q_{it_i} = (1 - q_{iy_i})/(K-1)$, and thus,

$$m_i = \frac{Kq_{iy_i} - 1}{K - 1}. \tag{6}$$

If we adopt this local strategy, $p = $ const means $\sum m_i = $ const again. We assume hereafter that the sum of margins is constant with this local optimal strategy.

Returning to margin maximization, in Criterion (2), there is nothing to do other than (5), because the value is constant under (5) and for a fixed p. We may equalize m_i to be $m_i = \sum m_i/N$. Criterion (1) is maximized by the same value of m_i's, that is,

$$m_i = \frac{Kp - 1}{K - 1}, \ \forall \ i. \tag{7}$$

That is, the solution is the same as that shown in [5] for which the minimum margin (1) is maximized in the oracle output scheme without condition (5).

The assignment maximizing Criterion (3) is quite different from the other two. It is directly connected to the performance of h.

From (6), we have

$$m_i \geq \theta \iff q_{iy_i} \geq \frac{1 + (K-1)\theta}{K},$$

equivalently,

$$\mathbb{I}(m_i \geq \theta) = \begin{cases} 1 & (q_{iy_i} \geq \frac{1+(K-1)\theta}{K}) \\ 0 & (\text{otherwise}) \end{cases}. \tag{8}$$

Therefore, if $\sum q_{iy_i} = Np < N(1 + (K-1)\theta)/K$, Criteria (3) cannot be maximized by making all m_i be equal unlike other two criteria. It is maximized by having as many i's satisfying $q_{iy_i} = (1 + (K-1)\theta)/K$ as possible and make $q_{iy_i} = 0$ for the remaining i's. Note that the last criterion is directly related to Theorem 1. Criteria (3) is the number of samples whose margin is greater than θ, and thus it corresponds to the term $P_S[margin(f, x, y) \leq \theta]$ in Theorem 1.

Theorem 2 (Maximization of Several Margin Criteria). *In K-class $(K > 2)$ problems and in the majority-voting ensemble of independent classifiers with a constant average accuracy p, (1) the minimum margin $(\min_i m_i)$ is maximized (a) when these classifiers vote evenly for every class other than the correct class in a training sample and (b) when the ratio of correct votes is the same over all the training samples, and (2) the average margin $\left(\frac{1}{N}\sum_{i=1}^{N} m_i\right)$ is maximized by (a) only. On the contrary, (3) the number of margins larger than a threshold $\theta > 0$ is maximized $\left(\sum_{i=1}^{N} \mathbb{I}(m_i \geq \theta)\right)$ (c) when some samples are discarded in such a way that for these samples no vote is casted to their correct classes and instead the number of samples satisfying the condition with the equality is increased.*

6   Diversity Measure and Margin

6.1   The Disagreement Measure

The disagreement measure is one of the measures of diversity [2]. According to the way as Tang $et\ al.$ [5] did, we extend the definition of disagreement measure from two-class to multi-class cases. Here, we extend only "disagreement measure," but the other diversity measures [5] are easily extended in a similar way.

First, we assume $K = 2$. In this two-class case, the disagreement measure of two classifiers, h_j and h_l, is defined as

$$dis_{j,l} = \frac{1}{N} \sum_{i=1}^{N} \mathbb{I}(c_{ji} \neq c_{li}),$$

and the diversity of the whole set of classifiers is represented as the mean of $dis_{j,l}$ over all possible pairs, that is,

$$dis = \frac{2}{L(L-1)} \sum_{j<l} dis_{j,l}. \tag{9}$$

Since $q_{it_i} = 1 - q_{iy_i}$ in $K = 2$, (9) can be written as

$$dis = \frac{2L}{N(L-1)} \sum_{i=1}^{N} q_{iy_i}(1 - q_{iy_i}). \tag{10}$$

We extend the disagreement measure from two-class case to multi-class case. As shown in (10), the disagreement measure represents the number of pairs of classifiers outputting different class labels, so it is naturally extended as

$$dis = \frac{L}{N(L-1)} \sum_{i=1}^{N} \sum_{k=1}^{K} q_{ik}(1 - q_{ik}). \tag{11}$$

It is easy to show that, under the condition of $\sum_{i=1}^{N} q_{iy_i} = Np$, this extended disagreement measure (11) is maximized by

$$q_{iy_i} = p, \quad q_{ik} = \frac{1-p}{K-1} \quad (\forall k \neq y_i), \quad i = 1, 2, \dots, N.$$

In other words, it is maximized when $\mathbf{q}_i$ is equal over all samples and the frequencies q_{ik} of classes other than the correct class are all equal. It is obvious in this case that $m_i = (Kp - 1)/(K - 1)$ for all i. This is the same as (7).

We note that such an equal distribution of probabilities maximizes the entropy too. Therefore, as a new diversity measure, it is natural to consider a generalized entropy, Rényi entropy, defined for $\mathbf{r} = (r_1, r_2, \dots, r_K)$ by

$$H_\alpha(\mathbf{r}) = \frac{1}{1-\alpha} \log \sum_{i=1}^{K} r_i^\alpha \quad \alpha \geq 0, \ \alpha \neq 1. \tag{12}$$

The Rényi entropy is maximized to $\log K$ by the equal distribution of $r_i = 1/K$, $\forall i$. In addition, for Shannon entropy ($\alpha \to 1$), we can show

$$H(r_1, r_2, \ldots, r_K) = (1 - r_1)H\left(\frac{r_2}{1-r_1}, \frac{r_3}{1-r_1}, \ldots, \frac{r_K}{1-r_1}\right) + H(r_1, 1 - r_1).$$

That is, given $q_{iy_i} = p$, entropy $H(q_{i1}, q_{i2}, \ldots, q_{iK})$ is maximized by the equal distribution of $\forall k \neq y_i$, $q_{ik} = (1 - p)/(K - 1)$. Thus, taking the sum of the entropy $H(\mathbf{q}_i)$ over $i = 1, 2, \ldots, N$, we can have a diversity measure. Moreover, since $x(1 - x) \simeq -x \log x$, the Shannon entropy measure behaves almost the same as the extended disagreement measure.

As another extreme case of Rényi entropy ($\alpha = \infty$), if $q_{iy_i} = \max (q_{i1}, q_{i2}, \ldots, q_{iK})$,

$$H_\infty(q_{i1}, q_{i2}, \ldots, q_{iK}) = -\log \max q_i = -\log q_{iy_i}.$$

Thus, if the correct class always wins the majority voting, we see

$$\sum_{i=1}^{N} q_{iy_i} = \sum_{i=1}^{N} \exp\left(-H_\infty(q_{i1}, q_{i2}, \ldots, q_{iK})\right).$$

This measure is insensitive to the vote distribution, but sensitive to which gained the largest number of votes in y_i and t_i, in other words, if m_i is positive or not. That is, this measure is most related to the accuracy of the ensemble classifier h.

Theorem 3 (Margin Maxmization and Disagreement Measure). *In K-class ($K > 2$) problems and in the majority-voting ensemble of independent classifiers with a constant average accuracy p, the maximization of the (extended) disagreement measure over the classifiers and Shannon entropy measure over votes of the classifiers, are coincident with the maximization of (1) the minimum margin ($\min_i m_i$) and (2) the average margin $\left(\frac{1}{N}\sum_{i=1}^{N} m_i\right)$.*

7   Experiment

7.1   Relationship Between Margins, the Disagreement Measure and Entropy

First, we investigated the correlation between several diversity measures and three maximum margin criteria presented so far.

For given values of p, N, L, we made a result table for N samples and L base classifiers as follows. We first generated the values of y_i ($i = 1, 2, \ldots, N$) at random according to the uniform distribution over $\mathcal{K} = \{1, 2, \ldots, K\}$, and then assigned the value of c_{ji} at random and uniformity in such a way that c_{ji} is identified with y_i with probability p. We made $T = 1000$ tables of size $L \times N$ and measured above statistics. In the first 500 tables, classes other than the correct class were chosen at random with probability of $(1 - p)/(K - 1)$,

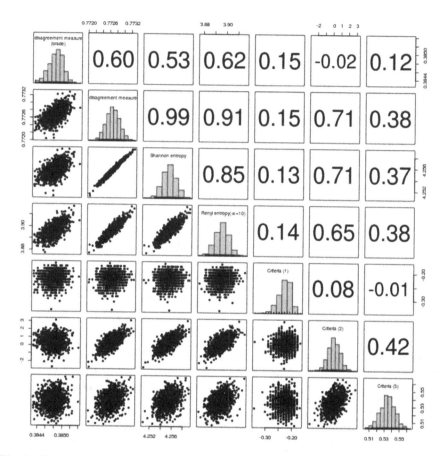

Fig. 3. Scatterplot matrix over four diversity measures in $K = 5$, $p = 0.26$, $N = 1000$, $L = 100$. From top, the disagreement measure in the oracle output, the disagreement measure in multi-class, Shannon entropy, Rényi entropy ($\alpha = 10$), three maximum margin criteria (from 5th row, $\min_i m_i$, $\sum m_i$ and $\sum \mathbb{I}(m_i > 0)$).

and in the second 500 tables, only $K - 2$ classes were chosen with probability $(1 - p)/(K - 2)$. In this setting, the value of Criterion (3) with $\theta = 0$ can be regarded as the accuracy of the ensemble classifier h.

From Fig. 3, we can see

1. As explained, even if the average accuracy p is less than $1/2$, as long as $p > 1/K$, the ensemble classifier h can have a higher accuracy p_h (in this example, $p_h \fallingdotseq 0.53$ for $p = 0.26$).
2. The extended multi-class disagreement measure has a larger positive correlation with p_h than the original oracle-based disagreement measure (0.38 vs. 0.12).
3. As explained, the extended disagreement measure is almost the same as the Shannon entropy with correlation coefficient of 0.99.

4. Rényi entropy with $\alpha = 10$ is comparable to Shannon entropy.
5. The minimum margin is almost independent of p_h.
6. Unfortunately, any diversity measure does not have a sufficient level of correlation with p_h.

8 Conclusion

We have investigated the relationship between a diversity measure and the margin in multi-class problems. Diversity measures are a popular tool to measure the degree to which independent base classifiers of accuracy slightly larger than $1/2$ are combined into a stronger ensemble classifier. However, they have been mainly defined in the oracle output scheme where the output of a base classifier is correct or wrong. We showed that this is not appropriate for the problems of more than two classes. In fact, a necessary condition to be satisfied by weak classifiers to produce a stronger ensemble classifier is weaken to $1/K$ (K is the number of classes) from $1/2$, and the meaning of positive margins changes. Therefore, we have extended the disagreement measure among base classifiers so as to be applicable to multi-class cases.

It is well known that maximizing the margin is reducing the generalization error of a classifier. However, only a few studies have been made on the relationship between the margin and several diversity measures, and they are all made in the oracle output scheme. Therefore, we studied the relationship in multi-class setting and revealed that in two of three criteria considered in this paper the margin is maximized when the conditional probabilities of classes other than the correct class are all the same, implying a possibility to use the entropy as another diversity measure. Using a generalized Rényi entropy and the extended disagreement measure, we confirmed theoretically and experimentally the effectiveness of these measures in the context of maximizing the margin.

Acknowledgment. This work was partly supported by JSPS KAKENHI Grant Numbers 25280079 and 15H02719.

References

1. Breiman, L.: Random forests. Mach. Learn. **45**(1), 5–32 (2001)
2. Ho, T.: The random subspace method for constructing decision forests. IEEE Pattern Anal. Mach. Intell. **20**(8), 832–844 (1998)
3. Kuncheva, L.I., Whitaker, C.: Measures of diversity in classifier ensembles and their relationship with the ensemble accuracy. Mach. Learn. **51**(2), 181–207 (2003)
4. Schapire, R.E., Freund, Y., Bartlett, P.L., Lee, W.S.: Boosting the margin: a new explanation for the effectiveness of voting methods. Ann. Stat. **26**(5), 1651–1686 (1998)
5. Tang, E., Suganthan, P., Yao, X.: An analysis of diversity measures. Mach. Learn. **65**(1), 247–271 (2006)

Fractional Programming Weighted Decoding for Error-Correcting Output Codes

Firat Ismailoglu[1]([✉]), I.G. Sprinkhuizen-Kuyper[2], Evgueni Smirnov[1], Sergio Escalera[3], and Ralf Peeters[1]

[1] Maastricht University, Department of Knowledge Engineering,
6211 LH Maastricht, The Netherlands
{f.ismailoglu,smirnov,ralf.peeters}@maastrichtuniversity.nl
[2] Radboud University Nijmegen, Donders Institute for Brain,
Cognition and Behaviour, Centre for Cognition, 6500 HE Nijmegen, The Netherlands
i.kuyper@donders.ru.nl
[3] University of Barcelona and Computer Vision Center, Barcelona, Spain
sergio@maia.ub.es

Abstract. In order to increase the classification performance obtained using Error-Correcting Output Codes designs (ECOC), introducing weights in the decoding phase of the ECOC has attracted a lot of interest. In this work, we present a method for ECOC designs that focuses on increasing hypothesis margin on the data samples given a base classifier. While achieving this, we implicitly reward the base classifiers with high performance, whereas punish those with low performance. The resulting objective function is of the fractional programming type and we deal with this problem through the Dinkelbach's Algorithm. The conducted tests over well known UCI datasets show that the presented method is superior to the unweighted decoding and that it outperforms the results of the state-of-the-art weighted decoding methods in most of the performed experiments.

1 Introduction

Reducing a multiclass classification problem to a series of binary classification problems has attracted an increasing attention in the machine learning community [1,2]. This allows one to use binary classification algorithms also for multiclass problems without a need for extending them. Besides, thanks to the reduction, one can take the merits of the ensemble learning methods, e.g. providing a better generalization ability [1]. The procedure of the reduction is governed by a decomposition schema. It determines the number of binary problems to be generated, thus the number of base classifiers to be learned, and together with how to group the original classes in two sets for each binary problem. The state-of-the-art decomposition schemas include One-vs-All, One-vs-One (pairwise coupling) and Error Correcting Output Codes (ECOC) [3]. Of these, ECOC has attracted the most attention primarily since one can *correct* the errors committed in the induced binary problems with ECOC. In addition, ECOC can reduce both bias and variance of the learning algorithm [4].

© Springer International Publishing Switzerland 2015
F. Schwenker et al. (Eds.): MCS 2015, LNCS 9132, pp. 38–50, 2015.
DOI: 10.1007/978-3-319-20248-8_4

As in the traditional ensemble learning methods, e.g. AdaBoost [6], combining the base hypotheses by crediting different weights to them is also of importance to ECOC. For ECOC, however, this should be considered from the perspective of *metric learning*, since it results in learning a task-specific distance function. Based on the task, different weighting methods have been proposed. Firstly, performance of the base classifiers vary, since the resulting binary classification problems are not equally solvable. Thus, in turn, we should not treat the outcome of the binary learners in the same way. The paper by Escalera et al. is built on this phenomenon [7]. As a second example, Smith et al. proposed to weight each base classifier for any class based on the way this classifier separates that class from the rest. They showed that the *class-separability* property improves the generalization performance of the final ensemble [8].

In contrast to the previous work we argue that weighting the base classifiers increases the *hypothesis margin* of query instances, meaning that instances will become closer to their true classes while increasing the distance to the rest. This indeed leads to an increase in the ensemble accuracy. In this paper, we address the issue of maximizing the hypothesis-margin of instances by means of classifier weighting using the Dinkelbach algorithm, which is a central algorithm in fractional programming [11].

The organization of the paper is as follows. In Sect. 2, we provide an insight about ECOC framework together with the prominent weighted decoding strategies. Section 3 details the proposed algorithm and its resemblance with the conventional weighted nearest neighbor algorithms. Section 4 presents the experiments conducted over several UCI datasets and the empirical comparison of the alike algorithms. Finally, with the Sect. 5, we conclude the work and give several avenues for future work.

2   Error-Correcting Output Codes with Weighted Decoding

In a multiclass classification problem, we are given a training set of pairs $\mathcal{D} = \{(x_i, y_i)\}_{i=1}^{N} \subset X \times Y \subset \mathbb{R}^d \times \{1, 2, ..., K\}$ where X is some instance space, and Y is a label space with $K > 2$. $\mathcal{D}$ is iid derived according to some unknown probability distribution p over $X \times Y$. The goal is to learn a multiclass classifier (polychotomizer) $\hat{F} : X \rightarrow \{1, 2, .., K\}$. One can directly learn $\hat{F}$ by a multiclass classification algorithm or extend a binary classification algorithm such that it also handles the multiclass case. Alternatively, we can make use of well-established decomposition strategy. That is, for a code matrix $M \in \{-1, 1\}^{K \times B}$, B number of replicas of the instance space are generated. The instance in each replica is then re-labeled with the aid of a column that is associated with that replica. At the end, we end up with B binary sets $\mathcal{D}_b = \{x_i, M_b(y_i)\}_{i=1}^{N}$, where $M_b : Y \rightarrow \{-1, 1\}$ s.t. $M_b(y) = M(y, b), \forall b \in \{1, ..., B\}$. This phase is also known as *(en)coding*.

For each binary set $\mathcal{D}_b$, we learn a binary classifier $f_b : X \rightarrow \{-1, 1\}$. When assigning a query instance x_q to one of the classes, the binary learners all run simultaneously, thus the outputs of the learners constructs a vector $f(x_q) = (f_1(x_q), ..., f_B(x_q))^T$. In the *decoding* phase, a decoder function $g : \mathbb{R}^B \times M(r, \cdot) \to \{1, ..., K\}$ is used to obtain $\hat{F}(x_q) = g(f(x_q))$, where $M(r, \cdot)$ is a row of M which is so-called a *class codeword*. Class codewords are unique to each class. Here, g can be cast as a nearest neighbor classifier, since it computes the distance between $f(x_q)$ and each codeword, then assigns x_q to the class whose codeword is nearest to $f(x_q)$.

Based on the output of the binary classification algorithm being employed, the decoder g is allowed to use different distance measure, which leads to different decoding types. If the base classifiers output crisp labels, i.e. -1 or 1 only, then the Hamming distance is usually concerned, which is basically the total number of elements that mismatch in the vector $f(x_q)$ and in the row of M under consideration [3]. Or, the base classifier is probabilistic and outputs the posterior probability of being positive class, then in this case, Euclidean metric is plugged into g. The decoding here is called *Euclidean decoding*. Note that in this case $M \in \{0, 1\}^{K \times B}$. When one is interested in *confidence-valued* binary classifiers $f_b : \mathcal{X} \to \mathbb{R}$, *loss-based decoding* can be considered [2]. In this case, the distance function used by g is defined as:

$$d(f(x), M(k, \cdot)) = \sum_{b=1}^{B} L(f_b(x) \cdot M(k, b)), \tag{1}$$

where L is a loss function. The common loss functions are linear: $L(\theta) = -\theta$, and exponential: $L(\theta) = e^{-\theta}$. We, however, use the exponential one in our work.

2.1   Weighted Decoding Background

We now extend our discussion of decoding to weighted decoding, where the weighted distance measure comes into play in the decoding process. Here, we define two forms of weighted distance between a class codeword and the vector of binary classifiers outputs, depending on Euclidean WE or loss-based distance WL respectively:

$$d_{WE}(f(x), M(k, \cdot)) = \sqrt{\sum_{b=1}^{B} W_{kb}(f_b(x) - M(k, b))^2}, \tag{2}$$

$$d_{WL}(f(x), M(k, \cdot)) = \sum_{b=1}^{B} W_{kb} L(f_b(x) \cdot M(k, b)), \tag{3}$$

with $W \in [0, 1]^{K \times B}$. Note that we learn a weight vector per class which will later be a row of the weight matrix and that when the entries of W are all 1, d_W reduces to the conventional (unweighted) distance. Moreover, we are not concerned with weighted Hamming distance, since in this case, a weight for a binary learner applies only when that learner makes a deterministic (0–1) error.

Weighted decoding methods differ in determining the W matrix. Escalera et al. [7,10] utilize the following fact to determine W: The binary problems that emerge during the decomposition of the multiclass problem have different characteristics, thus they are not equally hard to solve with the same classification algorithm. It follows naturally that the performance of the binary classification algorithm over the resulting binary problems vary in a great range. Therefore, it is wise not to trust the output of the binary learners equally. Additionally, since a binary learner is originally trained to separate positive and negative super (at the same time artificial) classes, its performance over the *real* classes is expected to be non-uniform. As a result, Escalera et al. propose that W_{kb} signifies the performance of the bth classifier on class k. This can be readily estimated as:

$$H_{kb} = \frac{1}{|S_k|} \sum_{x \in S_k} [\![f_b(x) = M(k,b)]\!], \tag{4}$$

where $|S_k|$ is the cardinality of class S_k. Having calculated H_{kb} entries, they compute weight matrix W normalizing H such that each row sums to 1, W_{kb} therefore becomes:

$$W_{kb} = \frac{H_{kb}}{\sum_{b=1}^{B} H(k,b)}. \tag{5}$$

The rationale behind normalizing the rows is to make the prior probability of considering each class for the final classification the same. Otherwise, the final classifier, i.e. g, would be biased towards the dominant class in the dataset. In our work, we also follow this approach.

Rather than assigning the weight matrix empirically, Zhang et al. [12] propose to optimize the weight matrix ensuring that the weighted distance between each instance in the training set and its true codeword is small compared to those obtained from the remaining codewords. Here, the emerging optimization problem is a convex linear programming problem, hence it can be solved globally.

According to Smith et al. [8], W_{kb} should stand for how well the classifier b separates the members of the kth class is; and to them, if such information is captured during the final step in ECOC, the ensemble accuracy will be higher. To this end, they estimate W_{kb} through the training set as follows:

$$W_{kb} = max\{0, \frac{1}{K_k} [\sum_{\substack{p \in S_k \\ q \notin S_k}} C_b(\mathbf{p})C_b(\mathbf{q}) - \sum_{\substack{p \in S_k \\ q \notin S_k}} \overline{C_b}(\mathbf{p})\overline{C_b}(\mathbf{q})]\}, \tag{6}$$

where C_b is a correction function which takes the value 1, if the bth classifier correctly classifies the instance; 0 otherwise. Conversely, $\overline{C_b}$ is just the complement of C_b defined as: $\overline{C_b}(x) = 1 - C_b(x)$; and K_k is a normalization constant that has the same task as in Escalera et al. [7,10]. We remark that this method is again an empirical estimation of the weight matrix, thus lacks of a theoretical optimization.

3   Fractional Programming Weighted Decoding for ECOC

Recall that in the previous section, we have emphasized that the decoder g acts as a nearest neighbor (NN) classifier. From this point of view, the idea of weighted decoding can be tied to that of weighted nearest neighbor; so the methods originally devised for weighted nearest neighbor can be adapted to weighted decoding case.

Parades and Vidal [13], in their weighted nearest neighbor (NN) work, propose a class-dependent weighted dissimilarity measure. They consider that NN accuracy improves, if the employed dissimilarity measure in the NN yields small values for the distances between points coming from the same class, while it returns high values for inter-class distances. With this in mind, the objective function that they aim at minimizing is:

$$J(W) = \frac{\sum_{x \in \mathcal{D}} d_W(x, x_{nn}^=)}{\sum_{x \in \mathcal{D}} d_W(x, x_{nn}^{\neq})}, \tag{7}$$

where $x_{nn}^=$ is the nearest neighbor of x in its own class and $x_{nn}^{\neq}$ is again nearest neighbor of x but in a different class. Also note that $x_{nn}^=$ and $x_{nn}^{\neq}$ are nothing but *nearest hit* and *nearest miss* in the RELIEF algorithm respectively.

Following to Parades and Vidal, we can adapt (7) to the weighted decoding problem. Our objective function then turns out to be:

$$J(W) = \frac{\sum_{x_i \in \mathcal{D}} d_W(f(x_i), M(y_i, \cdot))}{\sum_{x_i \in \mathcal{D}} d_W(f(x_i), M(y_i^{\neq}, \cdot))}, \tag{8}$$

subject to the constraints $\sum_b w_{kb} = 1 \; \forall k$, where $y_i^{\neq}$ is the *most confused class* for the generic instance x_i. By the term most confused class, we mean that it is not the true class of x_i, but its associated codeword is nearest to $f(x_i)$ considering the other class codewords. Besides, in our case d_W can be instantiated with both d_{WE} and d_{WL}. Here, minimizing (8) can be also interpreted as maximizing the hypothesis-margin of the instances, since in the case of nearest neighbor it is generally defined as:

$$\rho(x_i) = d_W(f(x_i), M(y_i^{\neq}, \cdot)) - d_W(f(x_i), M(y_i, \cdot)). \tag{9}$$

The reader is referred to [9,14] for more information. Note that $f(x)$ in (8) and (9), is a transformation of the instance x to the space where the decoding takes place. One normally performs this transformation only for the test instances running the binary classifiers over them. However, in order to compute the weight matrix, we also need to transform the training instances to this (decoding) space. To do so, after learning the binary classifiers either from a bootstrap or from the entire training set, we let them run over the full training set. As a result, each training instance, x_i, is represented by a vector: $f(x_i) = (f_1(x_i), f_2(x_i), ..., f_B(x_i))^T$. That is, x_i becomes $f(x_i)$.

The objective function (8) is of the *fractional programming* type, which is generally defined as:

$$\min_{z \in Z} r(z), where \ r(z) = \frac{s(z)}{t(z)}, \tag{10}$$

with s and t which are real-valued functions on some feasible set Z, and $t(z) > 0$, $\forall z \in Z$. Among the methods to tackle (10), the most common is the parametric method which reduces (10) to the following parametric problem:

$$\min_{z \in Z} r_\lambda(z), where \ r_\lambda(z) = s(z) - \lambda t(z), \lambda \in \mathbb{R}. \tag{11}$$

Assuming (11) has at least one optimal solution for each $\lambda \in \mathbb{R}$, it is proved that every optimal solution to (11) is also optimal for (10) [15]. To determine $\lambda \in \mathbb{R}$, we use the Dinkelbach's algorithm shown in Algorithm 1.

Algorithm 1. The Dinkelbach's Algorithm

Step 0: Let $z^0 \in Z$, $\lambda^1 = s(z^0)/t(z^0)$, and $l = 1$
Step 1: Find an optimal solution $z^l \in Z$ of $min\{s(z) - \lambda^l t(z)\}$.
Step 2: Let $\lambda^{l+1} = s(z^l)/t(z^l)$. If $\lambda^{l+1} = \lambda^l$ then z^l is an optimal solution for $r(z)$, λ^l is the optimal value, and STOP, else set $l = l + 1$ and go to **Step 1**.

When solving the cost function (8) via the Dinkelbach's Algorithm as such, we generate a sequence of auxiliary problems of the form as in (11). The sequence of solutions of the auxiliary problems converges to a solution of the fractional program in question. Considering our problem, handling such subproblems differs in the weighted distance measure, as defined in (2) or (3). A general framework on how to tackle the fractional program in the weighted ECOC scenario is given in Sect. 3.1.

3.1 The FP_Weighted Decoding Algorithm

Denoting our cost function (8) $J(W) = s(W)/t(W)$ subject to the same constraints, the auxiliary problems encountered in the Dinkelbach's Algorithm are of the following type:

$$\min_{W \in \mathcal{W}} \{s(W) - \lambda^l t(W)\}, \tag{12}$$

where $\mathcal{W}$ is the feasible set, $\lambda^l \in \mathbb{R}$ is the (fixed) parameter at the iteration step l, $s(W) = \sum_{x_i \in \mathcal{D}} d_W(f(x_i), M(y_i, \cdot))$, and $t(W) = \sum_{x_i \in \mathcal{D}} d_W(f(x_i), M(y_i^{\neq}, \cdot))$. With this current form of the objective function, the solution to (12), W^*, will be quite sparse. In other words, only a very few binary learners will be associated with non-zero weights for per class. Therefore, outputs of the vast majority of the binary learners will be ignored, which in fact will lead to a shrinkage for the code matrix, so the error-correcting property of ECOC will disappear. In order to avoid this, we add the negative entropy regularization term to (12). The modified subproblem is defined as:

$$\min_{W \in \mathcal{W}} \{s(W) - \lambda^l t(W) + h \sum_{k=1}^{K} \sum_{b=1}^{B} w_{kb} \log w_{kb}\}, \quad s.t. \sum_{b} w_{kb} = 1 \; \forall k. \quad (13)$$

The coefficient $h \geq 0$ governs the relative importance of the regularization term compared with the original subproblem. The larger this coefficient is, the more equal the weights become, as the negative entropy function gets its minimum value for equal weights. We also point out that adding a regularization term to the auxiliary problems in the Dinkelbach's Algorithm does not disturb the convergence of the solutions, provided that the added term is convex [16]. Finally, to convert the above constrained optimization problem to an unconstrained one, we introduce the Lagrange multipliers λ_k, which yields the following:

$$\min_{W \in \mathcal{W}} \{s(W) - \lambda^l t(W) + h \sum_{k=1}^{K} \sum_{b=1}^{B} w_{kb} \log w_{kb} + \sum_{k=1}^{K} \lambda_k \left(1 - \sum_{b=1}^{B} w_{kb}\right)\}. \quad (14)$$

We now define X_{kb}, based on which distance function is considered, as follows:

$$X_{kb} = \begin{cases} \sum_{x \in S_k} (f_b(x) - M(k,b))^2 - \lambda^l \sum_{x \notin S_k \wedge y_i^{\neq} = k} (f_b(x) - M(k,b))^2, & \text{if } d_{WE} \\ \sum_{x \in S_k} L(f_b(x) \cdot M(k,b)) - \lambda^l \sum_{x \notin S_k \wedge y_i^{\neq} = k} L(f_b(x) \cdot M(k,b)), & \text{if } d_{WL}. \end{cases} \quad (15)$$

Note that in either case X_{kb} is not a function of w_{kb}. Plugging (15) into (14) yields:

$$\min_{W \in \mathcal{W}} \{\sum_{k=1}^{K} \sum_{b=1}^{B} (w_{kb} X_{kb} + h w_{kb} \log w_{kb}) + \sum_{k=1}^{K} \lambda_k \left(1 - \sum_{b=1}^{B} w_{kb}\right)\}. \quad (16)$$

Interestingly, the resulting unconstrained problem (16), is of the same form as the one encountered in [17], in which the authors are concerned with obtaining a weight vector per cluster so that the weighted distance between each instance and its cluster center will be minimum compared to those obtained from the remaining cluster centers. Naturally, the proposed solution to (16) is similar to the one in [17]. We shall briefly derive the optimal solution to the above unconstrained problem; whereas a more detailed derivation can be found in [17].

Observe a negative correlation between X_{kb} and w_{kb} in (16). Regardless of the distance measure considered, X_{kb} consists of two competing terms. The first term can be read as an estimate of the total amount of error that the bth classifier has made over the members of the kth class, while the second term quantifies how the bth classifier discriminates the kth class from the others. Thus, the larger the first term is, the more we *punish* the bth classifier reducing the corresponding entry w_{kb}, and in the cases where the second term is large, we *reward* the bth classifier increasing w_{kb}.

To solve (16), we take the partial derivative of it with respect to w_{kb} and λ_k; then set these derivations equal to zero. This yields following equations respectively:

$$X_{kb} + h \log w_{kb} + h - \lambda_k = 0, \tag{17}$$

$$1 - \sum_{b=1}^{B} w_{kb} = 0. \tag{18}$$

From Eq. (17), we obtain $w_{kb} = \frac{exp(-X_{kb}/h)}{exp(1-\lambda_k/h)}$. Substituting this equation in (18); then solving the resulting equation w.r.t. λ_k, leads: $\lambda_k = -h \log \sum_{b=1}^{B} exp((-X_{kb}/h) - 1)$. Finally, plugging this expression into the solution of (17) takes us to the optimal w_{kb}:

$$w_{kb}^* = \frac{exp(-X_{kb}/h)}{\sum_{b=1}^{B} exp(-X_{kb}/h)}. \tag{19}$$

Embedding the above update rule into the Dinkelbach's Algorithm, we define Algorithm 2.

Algorithm 2. FP_Weighted Decoding

Require: Training set $\mathcal{D} \subset \mathbb{R}^d \times \{1, 2, ..., K\}$, code matrix $M \in \{0, 1\}^{K \times B}$ or $M \in \{-1, 1\}^{K \times B}$, binary classification algorithm $\mathcal{A}$, stopping threshold ξ, regularization param. h.

Output: $W \in [0, 1]^{K \times B}$ s.t. $\sum_b w_{kb} = 1 \; \forall k$.

    Step 1: $\forall b$ relabel $\mathcal{D}$ considering $M(\cdot, b)$, and learn a binary classifier f_b with $\mathcal{A}$.
    Step 2: $\forall b$ run f_b over each $x_i \in \mathcal{D}$, so that $x_i \rightarrow f(x_i) = (f_1(x_i), ..., f_B(x_i))^T$.
    Step 3: Let $W^0 \in \mathcal{W}$, $\lambda^1 = s(W^0)/W(z^0)$, and $l = 1$.
    Step 4: $\forall k$ and $\forall b$, $w_{kb}^* = \frac{exp(-X_{kb}/h)}{\sum_{b=1}^{B} exp(-X_{kb}/h)}$
    with X_{kb} as in (15).
    Step 5: Let $\lambda^{l+1} = s(W^l)/t(W^l)$. If $|\lambda^{l+1} - \lambda^l| \leq \xi$, then output W^l and STOP, else set $l = l + 1$ and go to **Step 4**.

---

4   Experiments

We have tested our proposed method on a number of UCI datasets summarized in Table 1. For these datasets, we have compared our method against the classical (unweighed) decoding and against two previously proposed weighted decoding methods: the one proposed by Escalera et al. [7], and the one by Smith et al. [8]. For brevity, we name the former Perf_Weighted (for performance weighted), and the latter Sep_Weighted (for separability weighted). For each induced binary problem, we have first learned the base classifiers using the entire training set, and then estimated W_{kb} using again the training set. A separate validation set could be used to provide a robust estimate for W_{kb} in some cases though; in our experiments we have observed that for more than half of the datasets the ECOC

Table 1. The characteristics of the used UCI datasets.

Dataset	# Train	# Test	# Attribute	# Class
Balance	625	-	4	3
Iris	150	-	4	3
Thyroid	215	-	5	3
Car	1728	-	6	4
Vehicle	846	-	18	4
Dermatology	366	-	34	6
Glass	214	-	9	7
Segmentation	2310	-	19	7
Zoo	101	-	16	7
Ecoli	336	-	8	8
Mfeat	2000	-	6	10
Optdigits	3823	1797	64	10
Pendigits	7494	3498	16	10

accuracy is higher considering the training set for this purpose. This holds true for all the weighting methods. Here, we only give the results obtained using the training set. Additionally, to facilitate a fair comparison among the weighting methods, we have applied different weighting methods, having learned the base classifiers from the training set. Thus we ensure that the difference in ECOC accuracy arises only from using different weights in the decoding process.

In our experiments,we have opted for the standard code matrix, i.e. the exhaustive code matrix [3], for the datasets with less than 10 classes. In this setting, the number of induced binary problems is equal to $2^{K-1} - 1$, where K is the number of unique classes in the dataset. For the datasets with 10 classes, we have used a code matrix which takes into account all balanced binary class partitions. That is, each column of the matrix coincides with a balanced binary class partition. By balanced binary class partition, we mean that the number of positive classes and the number of negative classes is equal after partitioning classes into two. In total, $\frac{K!}{2(\frac{K}{2}!)^2}$ class partitions are obtained in this way, which is in fact the total column number in the matrix used. The related discussion can be found in [18]. The reason to choose such a code matrix is that these matrices hold the property of having equidistant rows. The proposed method works best with the code matrix with this property, because determining the *most confused class*, otherwise, would not only depend on the base classifiers performances.

Tables 2 and 3 show the results obtained considering the (weighted) euclidean distance and the (weighted) loss-based distance respectively. Excluding the two datasets Pendigits and Optdigits, where the training set and the test set are given separately, the results shown in the these tables are based on the average of the 10-fold cross validation runs, ensuring that each method uses the same folds for training and uses the same fold for testing in each run. For Pendigits

Table 2. (Un)weighted Euclidean decoding results with the base classifier logistic regression.

Dataset	Unweighted	FP_Weighted	Perf_Weighted	Sep_Weighted
Balance	86.66	87.23	91.32	**91.44**
Thyroid	**95.8**	**95.8**	**95.8**	**95.8**
Iris	96.66	**98**	97.33	97.33
Car	81.13	80.09	**82.86**	82.63
Vehicle	79.29	**80.28**	79.25	79.49
Glass	60.87	**62.92**	58.99	58.92
Dermatology	95.49	**96.15**	**96.15**	**96.15**
Segment	83.02	**84.04**	83.9	83.72
Ecoli	82.15	81.86	**82.45**	**82.45**
Zoo	**88.11**	**88.11**	**88.11**	**88.11**
Mfeat	63.61	71.12	70.61	**72.27**
Pendigits	83.3	**86.96**	84.27	84.24
Optdigits	92.59	**93.65**	93.09	93.21

and Optdigits, the corresponding figures on the tables are obtained from their test set.

As binary classification algorithm, we have employed logistic regression and SVM with RBF kernel provided in LIBSVM [19], for the cases of weighted Euclidean and weighted loss-based decodings respectively. Also, the considered loss function is the exponential loss function which is defined as $L(\theta) = e^{-\theta}$. We note that the linear loss can not be used with our method, since the denominator in the objective function (8) could be negative in that case.

4.1 Evaluation of the Results

In Table 2 (resp. Table 3), the average ranks are 3.39, 2, 2.38 and 2.23 (resp. 3.58, 1.92, 2.37 and 2.13) for the unweighted, FP_Weighted, Perf_Weighted and Sep_Weighted decodings respectively. For both tables, the observed difference between the average ranks is significant according to the Freidman Test [20]. This implies that using different decoding strategies leads to different ECOC accuracies. Proceeding with the two-tailed Nemenyi test for pairwise comparisons, only the difference between the average rank of the proposed method and that of unweighted decoding is significant at $p = 0.05$. At $p = 0.1$ significance level, however, the weighted decodings are all superior to the unweighted decoding, taking the average rank differences into account. We also report that the conducted tests are not sufficient to make a pairwise comparison between the weighted decoding methods. These statements hold true for both Euclidean and loss-based decodings.

Table 3. (Un)weighted loss-based decoding with the base classifier SVM (with RBF).

Dataset	Unweighted	FP_Weighted	Perf_Weighted	Sep_Weighted
Balance	90.5	**94.46**	90.61	90.73
Thyroid	93.6	**93.99**	**93.99**	**93.99**
Iris	72	92.66	**93.33**	**93.33**
Car	79.17	79.88	84.05	**84.33**
Glass	47.14	**47.63**	46.64	46.72
Dermatology	93.47	**94.85**	93.85	93.85
Segment	78.71	83.93	86.85	**87.06**
Ecoli	40.39	50.54	**68.23**	65.9
Zoo	53.44	72.62	78.87	**79.06**
Mfeat	42.56	**42.77**	42.38	42.38
Pendigits	74.49	**76.44**	75.9	75.02
Optdigits	94.6	**95.9**	94.6	94.6

Taking magnitude of the differences into account per dataset, we reach the following conclusions. Table 2 suggests that introducing weights for the decoding process does not substantially improve the ECOC accuracy when the Euclidean distance is of interest. In fact, the improvement on the accuracy is at most 2 %, excluding mfeat dataset where there is an improvement of 8 % with the proposed method and that of 9 % with Sep_Weighted decoding. Nevertheless,

Table 4. The optimal regularization parameters for the proposed FP_Weighted algorithm.

Dataset	Euclidean decoding	Loss-based decoding
Balance	1	10
Iris	1	25
Thyroid	1	10
Car	100	100
Vehicle	10	-
Dermatology	less than 100	10
Glass	10	100
Segmentation	10	70
Zoo	less than 100	30
Ecoli	less than 100	70
Mfeat	30	50
Optdigits	30	10
Pendigits	120	30

weights come in more handy in loss-based decoding. For Iris, Segment, Ecoli and Zoo datasets the increase in the accuracy lies in the range of 9 % to 25 % comparing to unweighed decoding. Our proposed method, however, does significantly worse than the other two weighting methods for the aforementioned datasets.

Finally, it should be also reported that the proposed method is keenly sensitive to the regularization parameter h being used. Recalling that as h grows, the distribution of the weight values converges to the uniform distribution, which is basically the unweighed decoding. Table 4 shows the optimal h values for the different datasets found via cross-validation.

5 Conclusion

We presented an algorithm which outputs an optimal weight matrix to be used in the decoding phase of ECOC. The presented method takes its root in weighted nearest neighbor algorithms and guarantees that the distance between a coded instance and its target codeword will be small, while the distance between that instance and its most confused codeword will be large. This approach boosts the hypothesis-margin of the instances, thus indeed leads to a significant increase in the accuracy obtained from the ECOC. The performance of the proposed method maximizes when the used code matrix has equidistant rows and the base classifiers are scoring classifiers.

In the case of euclidean decoding, our proposed method achieves the best accuracy on 7 out of 11 datasets comparing with unweighted decoding as well as the other two state-of-the-art weighted decoding algorithms. Also, in the case of loss-based decoding, it exhibits a better performance than the unweighted case, but is inferior to the these weighted decoding methods. Finally, the future work will focus on applying different regularization terms to the objective function and on applying the proposed algorithm over a higher diversity of ECOC coding designs and different application domains.

References

1. Kuncheva, L.I.: Combining Pattern Classifiers: Methods and Algorithms. Wiley, Hoboken (2004)
2. Allwein, E.L., Schapire, R.E., Singer, Y.: Reducing multiclass to binary: a unifying approach for margin classifiers. J. Mach. Learn. Res. 1, 113–141 (2001)
3. Dietterich, T.G., Bakiri, G.: Solving multiclass learning problems via error-correcting output codes. J. Artif. Intell. Res. (JAIR) 2, 263–286 (1995)
4. Kong, E.B., Dietterich, T.G.: Error-correcting output coding corrects bias and variance. In: ICML, pp. 313–321 (1995)
5. Guruswami, V., Sahai, A.: Multiclass learning, boosting, and error-correcting codes. In: 12th Annual Conference Computational Learning Theory, Santa Cruz, California, pp. 145–155 (1999)
6. Schapire, R.E., Freund, Y.: Boosting: Foundations and Algorithms, vol. 1. MIT Press, Cambridge (2012)

7. Escalera, S., Pujol, O., Radeva P.: Loss-weighted decoding for error-correcting output codes. In: International Conference on Computer Vision Theory and Applications, Madeira, Portugal (2008)

8. Smith, R.S., Windeatt, T.: Class-separability weighting and bootstrapping in error-correcting output code ensembles. In: El Gayar, N., Kittler, J., Roli, F. (eds.) MCS 2010. LNCS, vol. 5997, pp. 185–194. Springer, Heidelberg (2010)

9. Crammer, K., Gilad-Bachrach, R., Navot, A., Tishby, N.: Margin analysis of the LVQ algorithm. In: Proceedings of 17th Conference on Neural Information Processing Systems (2002)

10. Escalera, S., Pujol, O., Radeva, P.: On the decoding process in ternary error-correcting output codes. IEEE Trans. Pattern Anal. Mach. Intell. **32**(1), 120–134 (2010)

11. Bajalinov, E.B.: Linear-Fractional Programming: Theory, Methods, Applications and Software, 1st edn. Kluwer Academic Publishers, New York (2003)

12. Zhang, X., Wu, J., Chen, Z., Lv, P.: Optimized weighted decoding for error-correcting output codes. In: IEEE International Conference on Acoustics, Speech and Signal Processing, Kyoto, Japan (2012)

13. Parades, R., Vidal, E.: A class-dependent weighted dissimilarity measure for nearest neighbor classification problems. Pattern Recogn. Lett. **21**(12), 1027–1036 (2000)

14. Sun, Y., Todorovic, S., Li, J., Wu, D.: Unifying the error-correcting output code AdaBoost within the margin framework. In: 22nd ICML, Bonn, Germany (2005)

15. Sniedovich, M.: Dynamic Programming Foundations and Principles, 2nd edn. CRC Press, USA (2011)

16. Gugat, M.: Prox-regularization methods for generalized fractional programming. J. Optim. Theory Appl. **99**(3), 691–722 (1998)

17. Domeniconi, C., Gunopulos, D., Ma, S., Yan, B., Al-Razgan, M., Papadopoulos, D.: Locally adaptive metrics for clustering high dimensional data. J. Data Min. Knowl. Discov. **14**(1), 63–67 (2007)

18. Smirnov, E., Moed, M., Kuyper, I.: Minimally-Sized Balanced Decomposition Schemes for Multi-class Classification. Ensembles in Machine Learning Applications. Springer, Berlin (2011)

19. Chang, C.C., Lin, C.J.: LIBSVM: a library for support vector machines. ACM Trans. Intell. Syst. Technol. **2**, 27:1–27:27 (2011)

20. Demsar, J.: Statistical comparisons of classifiers over multiple data sets. JMLR **7**, 1–30 (2006)

Instance-Based Decompositions of Error Correcting Output Codes

Firat Ismailoglu[1]([⊠]), Evgueni Smirnov[1], Nikolay Nikolaev[2], and Ralf Peeters[1]

[1] Department of Knowledge Engineering, Maastricht University,
P.O.BOX 616, 6200 MD Maastricht, The Netherlands
{f.ismailoglu,smirnov,ralf.peeters}@maastrichtuniversity.nl
[2] Department of Computing, Goldsmiths College, University of London,
London SE14 6NW, UK
n.nikolaev@gold.ac.uk

Abstract. This paper proposes *instance* decomposition schemes (IDSs) for mapping multi-class classification tasks into a series of binary classification tasks. It demonstrates theoretically that IDSs can handle two main problems of the *class* decomposition schemes: the problem of difficult binary classification tasks and the problem of positive error correlation of the binary classifiers. The experiments show that IDSs can outperform standard ECOC class decompositions.

1 Introduction

A *class* decomposition scheme allows mapping a multi-class classification task into a set of binary classification tasks [5]. In this way the multi-class classification task can be solved by a set of binary classifiers that correspond to the set of binary classification tasks identified. In general, a class decomposition scheme consists of several class partitions. Any class receives a code word that indicates the class set including that class for each partition. The scheme is applied in two stages [5,10]: encoding and decoding. During the encoding stage we first generate binary classification tasks according to the decomposition scheme and then train a binary classifier for each problem. During the decoding stage we first apply the binary classifiers for a test instance to generate the instance code word and then assign a class to that instance with the closest code word.

The most successful family of class decomposition schemes is that of error-correcting output codes (ECOC) [5]. Due to redundancy of the binary partitions, ECOC schemes can significantly improve generalization performance on multi-class classification tasks. However, two problems may occur:

(P1) the problem of difficult binary classification tasks; and
(P2) the problem of positive error correlation of the binary classifiers.

Problem P1 means that some binary classification tasks are difficult for the binary classifiers. As a result the code words of the test instances can contain binary classification errors. ECOC class decomposition schemes can correct these

F. Schwenker et al. (Eds.): MCS 2015, LNCS 9132, pp. 51–63, 2015.
DOI: 10.1007/978-3-319-20248-8_5

errors if their number is smaller than $\lfloor \frac{H_{min}-1}{2} \rfloor$ where H_{min} is the minimal Hamming distance between class code words. Otherwise, the correction fails which results in instance misclassification.

Problem P2 implies that the binary classifiers can err simultaneously when the code words of test instances are being formed. This increases the number of errors in the instance code words that can exceed the limit of $\lfloor \frac{H_{min}-1}{2} \rfloor$. If this happens, the instances may be misclassified. We note that instance misclassification can occur even if we have good binary classifiers as long as they positively error correlated.

Several approaches to handle the problem of difficult binary classification tasks were proposed for class decomposition schemes. We separate these approaches into several categories: (a) approaches that simply remove the partitions of the difficult binary problems [15]; (b) approaches that assign weights to partitions to maximize the Hamming distance between the classes in a class decomposition scheme [1]; (c) approaches that alter the partitions in order to simplify the binary classification tasks encoded [13,16]; and (d) approaches that finely tune the mappings of the classes and data into class decomposition schemes [4,14]. We note that none of these approaches employs information about the binary-classifier error correlation. Thus, if the error correlation is reduced, this is primarily due to improving the binary classifiers.

This paper proposes a new type of decomposition schemes, namely *instance* decomposition schemes (IDSs). They aim at handling the problem of difficult binary classification tasks and the problem of positive error correlation of the binary classifiers. An IDS scheme consists of several instance partitions. Any instance has a code word that indicates the instance set including that instance for each partition. Instance classification assumes two IDSs: encoding and decoding. During the encoding stage we first generate binary classification tasks using the encoding IDS, then train a binary classifier for each task, and, finally, we learn the decoding IDS through the binary classifiers. During the decoding stage we first apply the binary classifiers for a test instance to generate the instance code word and then we assign the class equal to the class of the instance with the closest code word in the decoding IDS.

In this paper we show that the problem of difficult binary classification tasks does not exist for IDSs. This is due to decoding IDSs that are learned w.r.t. the binary classifiers. In addition, we show that when the binary instances err in a similar way over the instances of different classes, IDSs have always error correcting capabilities in contrast with the ECOC class decomposition schemes. Thus, in this case IDSs represent error-correcting output codes.

The paper is organized as follows. Section 2 formalizes the classification task. IDSs are introduced and discussed in Sects. 3 and 4, respectively. Experiments are given in Sect. 5. Section 6 concludes the paper.

2   Classification Task

Let X be an instance space, Y be a class set of size K, and p be an unknown probability distribution over $X \times Y$. Training data T is a multi-set of L labeled instances $(x_l, y_l) \in X \times Y$ $(l \in 1..L)$ iid drawn from p. Given training data T and test instance $x \in X$, the classification task CT is to provide an estimate $\hat{y} \in Y$ of the class of x according to p. We note that if $K = 2$, the classification task is a *binary classification task BCT*. If $K > 2$, the classification task is a *multi-class classification task MCT*.

3 Instance-Based Decomposition Schemes

This section introduces instance-base decomposition schemes for mapping any multi-class classification task into a set of binary classification tasks. Section 3.1 formalizes *instance decomposition schemes and coding matrices*. Section 3.2 provides a detailed explanation of *the encoding and decoding stages*.

3.1 Instance Decomposition Schemes

Consider a multi-class classification task MCT with a class set Y of size $K > 2$. To decompose MCT into M binary classification tasks BCT_m $(m \in 1..M)$ we introduce the notion of a *binary instance partition* in Definition 1.

Definition 1 (Binary Instance Partition). *Given data T, the set $P(T)$ is said to be a binary instance partition of T iff $P(T)$ consists of two non-empty sets T^- and T^+ such that $T^- \cup T^+ = T$ and $T^- \cap T^+ = \emptyset$.*

The sets T^- and T^+ of a binary instance partition $P(T)$ are called the negative set and the positive set of $P(T)$, respectively.

Definition 2. *The label set $Y_{P(T)}$ of a binary instance partition $P(T)$ is defined equal to $\{-1, +1\}$, where -1 is the label of the negative set T^- of $P(T)$ and $+1$ is the label of the positive set T^+ of $P(T)$.*

Definition 1 allows us to introduce *instance decomposition schemes*. An instance decomposition scheme describes how to decompose a multi-class classification task MCT into M binary classification tasks BCT_m $(m \in 1..M)$, as given in Definition 3.

Definition 3 (Instance Decomposition Scheme). *Given a multi-class classification task MCT and a positive integer M, an instance decomposition scheme of MCT is a set $SP(T)$ of M different binary instance partitions $P_m(T)$ for $m \in 1..M$.*

Any instance decomposition scheme has a coding matrix (see Fig. 1).

Definition 4. *The coding matrix of an instance decomposition scheme $SP(T)$ is a binary $L \times M$ matrix W iff for any $l \in 1..L$ and $m \in 1..M$:*

$$W_{l,m} = \begin{cases} -1 \in Y_{P_m(T)} & \text{if } (x_l, y_l) \in T_m^-; \\ +1 \in Y_{P_m(T)} & \text{if } (x_l, y_l) \in T_m^+. \end{cases}$$

A row $W_{l,*}$ in the coding matrix W corresponds to a particular labeled instance $(x_l, y_l) \in T$ for any $l \in 1..L$ and it forms the *code word* of that instance. We note that the meaning of any two bits W_{l,m_1} and W_{l,m_2} in an instance code word is different for $m_1 \neq m_2$, since they correspond to different instance partitions. In addition it is worth mentioning that the instance code words are not restricted: the code words $W_{l_1,*}$ and $W_{l_2,*}$ of two different instances $(x_{l_1}, y_{l_1}), (x_{l_2}, y_{l_2}) \in T$ can be either the same or different. In this context we introduce the notion of the row distance set.

Definition 5. *Given an coding matrix W, the row distance set $D_W(l_1, l_2)$ for any two $l_1, l_2 \in 1..L$ is the set of indices $m \in 1..M$ such that $W_{l_1,m} \neq W_{l_2,m}$.*

The Hamming distance between any two instances $(x_{l_1}, y_{l_1}), (x_{l_2}, y_{l_2}) \in T$ in the coding matrix W is the size of the row distance set $D_W(l_1, l_2)$.

A column $W_{*,m}$ in the coding matrix W corresponds to a particular binary data partition $P_m(T)$ for any $m \in 1..M$ and it forms the *code word* of that partition. By Definition 3 any two partition code words W_{*,m_1} and W_{*,m_2} are different if $m_1 \neq m_2$. In this context we introduce the notion of the column distance set.

Definition 6. *Given an coding matrix W, the column distance set $D_W(m_1, m_2)$ for any two $m_1, m_2 \in 1..M$ is the set of the indices $l \in 1..L$ such that $W_{l,m_1} \neq W_{l,m_2}$.*

The Hamming distance between any two binary instance partitions $P_{m_1}(T), P_{m_2}(T) \in SP(T)$ in the coding matrix W is the size of the column distance set $D_W(m_1, m_2)$.

3.2   Encoding and Decoding

To solve a multi-class classification task MCT by employing instance decomposition schemes we need to pass two stages, *encoding* and *decoding*, that employ two instance decomposition schemes $SP^e(T)$ and $SP^d(T)$, respectively. During the encoding stage we use the coding matrix W^e of the *encoding* instance decomposition scheme $SP^e(T)$ (see Fig. 1). The matrix W^e is employed: (1) for representing the multi-class classification task MCT via a set of binary classification tasks BCT_m ($m \in 1..M$), and (2) for training binary classifiers for those problems. During the decoding stage we use the coding matrix W^d of the *decoding* instance decomposition scheme $SP^d(T)$. The matrix W^d is employed to decode the predictions provided by the binary classifiers to estimate the true class of an instance to be classified (see Fig. 1).

During the encoding stage we first generate for any $m \in 1..M$ a binary classification task BCT_m using the *encoding* matrix W^e. Any BCT_m is determined by code word $W^e_{*,m}$ of partition $P^e_m(T) \in SP^e(T)$. The data T_m of BCT_m is formed in $X \times Y_{P^e_m(T)}$. More precisely, any instance $(x_l, y_l) \in T$ is transformed to a new instance $(x_l, W^e_{l,m}) \in T_m$. Once the binary classification tasks BCT_m have been

$$W^e = \begin{pmatrix} +1 & +1 & +1 & +1 & +1 & +1 & +1 \\ +1 & +1 & +1 & +1 & +1 & +1 & +1 \\ -1 & -1 & -1 & -1 & +1 & +1 & +1 \\ -1 & -1 & -1 & -1 & +1 & +1 & +1 \\ -1 & -1 & -1 & -1 & +1 & +1 & +1 \\ -1 & -1 & +1 & +1 & -1 & -1 & +1 \\ -1 & +1 & -1 & +1 & -1 & +1 & -1 \\ -1 & +1 & -1 & +1 & -1 & +1 & -1 \end{pmatrix} \qquad W^d = \begin{pmatrix} +1 & +1 & +1 & -1 & -1 & -1 & +1 \\ -1 & -1 & -1 & +1 & +1 & +1 & +1 \\ -1 & -1 & +1 & +1 & -1 & +1 & -1 \\ +1 & +1 & +1 & -1 & +1 & +1 & +1 \\ -1 & -1 & -1 & +1 & +1 & +1 & +1 \\ +1 & -1 & +1 & +1 & -1 & -1 & +1 \\ -1 & +1 & -1 & -1 & +1 & +1 & -1 \\ -1 & -1 & -1 & +1 & +1 & +1 & +1 \end{pmatrix}$$

Fig. 1. Left: Coding matrix W^e of an encoding instance decomposition scheme $SP^e(T)$ for 4 classes. W^e is initialized according to the Exhaustive ECOC class decomposition. The first two rows correspond to the two instances of class y_1; The next three rows correspond to the three instances of class y_2; and so on. **Right:** Coding matrix W^d of the decoding instance decomposition scheme $SP^d(T)$ learned from W^e.

set, we train a binary classifier $h_m : X \to Y_{P_m^e(T)}$ for each BCT_m. The binary classifiers h_m form an ensemble classifier $h : X \to Y$ equal to $\{h_m\}_{m \in 1..M}$.

During the decoding stage, given a test instance $x \in X$ and an ensemble classifier h, we decode the predictions of the binary classifiers $h_m \in h$ to form a class estimate of the class for x. For that purpose we employ the coding matrix W^d of the *decoding* instance decomposition scheme $SP^d(T)$. We first estimate the code word w of instance x. The m-th element w_m of the code word w equals the label $h_m(x)$ estimated by the binary classifier $h_m \in h$ for x. Once the code word w of instance x has been formed, we temporarily add w to matrix W^d with index $L+1$ (i.e., $W^d_{L+1,*}$ equals w). Then we employ Hamming class decoding by computing the index neighbor set $N(x)$ for x. The set $N(x)$ consists of the indices $l \in 1..L$ of the instances $(x_l, y_l) \in T$ which code words $W^d_{l,*}$ have minimal Hamming distance to the code word w of instance x:

$$N(x) = \underset{l \in 1..L}{\mathrm{argmin}}(|D_{W^d}(l, L+1)|).$$

The set $\hat{Y}$ of estimates of the true class for the instance x is chosen to be the set of classes $y \in Y$ that have the maximal number of indices in $N(x)$; i.e., it is equal to $\underset{y \in Y}{\mathrm{argmax}}(|\{l \in N(x)|y_l = y\}|)$. The final class estimate $\hat{y}$ of instance x is chosen randomly among the classes in $\hat{Y}$. Note that this rule comprises the two possible cases: $|\hat{Y}| = 1$ and $|\hat{Y}| > 1$.

4   Initialization and Properties

The generalization performance of the ensemble h is sensitive to the initialization of instance decomposition schemes $SP^e(T)$ and $SP^d(T)$. Below we propose a simple procedure for initializing the coding matrices W^e and W^d of these schemes.

Definition 7. (Initialization Matrix Procedure)

(1) The encoding matrix W^e of $SP^e(T)$ is initialized such that:

$$(\forall l_1, l_2 \in 1..L)(W^e_{l_1,*} = W^e_{l_2,*} \leftrightarrow y_{l_1} = y_{l_2}).$$

(2) The decoding matrix W^d of $SP^d(T)$ is initialized such that:

$$(\forall l \in 1..L, m \in 1..M)(W^d_{l,m} = h_m(x_l)).$$

By Definition 7 the encoding matrix W^e is initialized such that the code words $W^e_{l_1,*}$ and $W^e_{l_2,*}$ of any two instances (x_{l_1}, y_{l_1}), $(x_{l_2}, y_{l_2}) \in T$ are equal iff the classes y_{l_1} and y_{l_2} are equal. This means that the instances of a class have the same code word and this word differs the code words of the instances of the remaining classes. Thus, the encoding matrix W^e is initialized according to the coding matrix W^0 of some *class* decomposition scheme (e.g., exhaustive/minimal/random ECOC etc. [6]).

By Definition 7 the coding matrix W^d of the *decoding* instance decomposition scheme $SP^d(T)$ is initialized using the predictions of binary classifiers $h_m \in h$. Thus, *the key feature of the initialization procedure is that we learn* $SP^d(T)$. We note that this is done by first training binary classifiers $h_m \in h$ using the *encoding* instance decomposition scheme $SP^e(T)$ and then by applying these classifiers. Hence, the coding matrix W^d of $SP^d(T)$ consists of only those instance code words that are achievable through binary classifiers $h_m \in h$. This implies that the problem of difficult binary classification tasks does not exist (if the binary classifiers are not random).

The decoding matrix W^d has to be well row-separated in terms of the Hamming distance [5]. This is due to the fact that W^d is used to decode binary predictions to form the final class estimate. The encoding matrix W^e however has to be both: row-separated and column-separated. The row-separation requirement comes from the fact that W^e is used to form W^d through the binary classifiers. The column-separation requirement comes from the fact that the binary classifiers need to be negatively error-correlated. We note that the usual assumption is that very different binary class labeling of two binary classification tasks may decrease such negative error correlation [5].

The minimal row Hamming distance in the coding matrix W^e of the *encoding* instance decomposition scheme $SP^e(T)$ is equal to the size of the minimal row-distance set $D_{W^e}(l_1, l_2)$ over any two instances (x_{l_1}, y_{l_1}), $(x_{l_2}, y_{l_2}) \in T$. By Definition 7 it follows that this distance is equal to the minimal row Hamming distance of coding matrix W^0 of the *class* decomposition scheme used for initializing the encoding matrix W^e.

To characterize the minimal row Hamming distance in the matrix W^d of the *decoding* instance decomposition scheme $SP^d(T)$ we note that the binary classifiers may err differently for any two instances. Hence, we introduce the error set for an instance.

Definition 8. *The error set $C_{W^e}(l)$ for an instance $(x_l, y_l) \in T$ is the set of indices $m \in 1..M$ of the binary classifiers $h_m \in h$, trained through the encoding matrix W^e, that err for that instance.*

Using error sets we compute the row distance set between any two instances in W^d.

Theorem 9. *The row distance set $D_{W^d}(l_1, l_2)$ for any two instances $(x_{l_1}, y_{l_1}), (x_{l_2}, y_{l_2}) \in T$ w.r.t. the decoding matrix W^d, if $y_{l_1} \neq y_{l_2}$, equals:*

$$\left[D_{W^e}(l_1, l_2) \setminus [[C_{W^e}(l_1) \setminus C_{W^e}(l_2)] \cup [C_{W^e}(l_2) \setminus C_{W^e}(l_1)]] \right] \cup$$

$$\left[D'_{W^e}(l_1, l_2) \cap [[C_{W^e}(l_1) \setminus C_{W^e}(l_2)] \cup [C_{W^e}(l_2) \setminus C_{W^e}(l_1)]] \right].$$

Proof. *The proof is by construction of the decoding matrix W^d in Definition 7.*

Example 10. Take row 1 (class y_1) and row 3 (class y_2) in W^e and W^d in Fig. 1. Then $D_{W^e}(1,3) = \{1,2,3,4\}$, $D'_{W^e}(1,3) = \{5,6,7\}$, $C_{W^e}(1) = \{4,5,6\}$, and $C_{W^e}(3) = \{3,4,5,7\}$. Thus, $D_{W^d}(1,3) = \{1,2,4,6,7\}$.

The minimal row Hamming distance in the decoding matrix W^d is equal to the size of the minimal set $D_{W^d}(l_1, l_2)$ over any two instances $(x_{l_1}, y_{l_1}), (x_{l_2}, y_{l_2}) \in T$. Due to non-uniform generalization performance of the binary classifiers there is no analytical way to express this distance in general. However, below we provide three particular cases when minimal row Hamming distance in the decoding matrix W^d can be derived.

Corollary 11. *For any two instances $(x_{l_1}, y_{l_1}), (x_{l_2}, y_{l_2}) \in T$, such that $y_{l_1} \neq y_{l_2}$, the row distance set $D_{W^d}(l_1, l_2)$ in the decoding matrix W^d equals the row distance set $D_{W^e}(l_1, l_2)$ of the encoding matrix W^e, if $C_{W^e}(l_1) \setminus C_{W^e}(l_2) = \emptyset$ and $C_{W^e}(l_2) \setminus C_{W^e}(l_1) = \emptyset$.*

Corollary 11 states that row distance sets $D_{W^e}(l_1, l_2)$ and $D_{W^d}(l_1, l_2)$ for instances $(x_{l_1}, y_{l_1}), (x_{l_2}, y_{l_2}) \in T$ stay equal if the two sets of the binary classifiers that err for those instances coincide. Thus, if this condition holds for any two instances, the minimal row Hamming distance in W^e equals that of the coding matrix W^d. By Definition 7 the minimal row Hamming distances in the encoding matrix W^e and the coding matrix W^0 of the class decomposition scheme used for initializing W^e coincide. Thus, the minimal row Hamming distance in W^d equals that of the coding matrix W^0.

Example 12. Take row 2 (class y_1) and row 4 (class y_2) in W^e and W^d in Fig. 1. Then $C_{W^e}(2) = \{1,2,3\}$, and $C_{W^e}(4) = \{1,2,3\}$. Thus, $D_{W^e}(2,4) = D_{W^d}(2,4) = \{1,2,3,4\}$.

Corollary 13. *For any two instances $(x_{l_1}, y_{l_1}), (x_{l_2}, y_{l_2}) \in T$, such that $y_{l_1} \neq y_{l_2}$, the row distance set $D_{W^d}(l_1, l_2)$ in the decoding matrix W^d equals $\{m \in 1..M\}$, if $C_{W^e}(l_1) \cap C_{W^e}(l_2) = \emptyset$ and $C_{W^e}(l_1) \cup C_{W^e}(l_2) = D'_{W^e}(l_1, l_2)$.*

Corollary 13 states that the row distance set $D_{W^d}(l_1, l_2)$ for instances (x_{l_1}, y_{l_1}), $(x_{l_2}, y_{l_2}) \in T$ is maximal (i.e., equal to $\{m \in 1..M\}$), if the two sets of the binary classifiers are disjointed and their union equals the compliment of $D_{W^e}(l_1, l_2)$. Thus, if this condition holds for any two instances, the minimal row Hamming distance in the decoding matrix W^d equals M.

Example 14. Take row 6 (class y_3) and row 7 (class y_4) in W^e and W^d in Fig. 1. Then $D'_{W^e}(6,7) = \{1, 4, 5\}$, $C_{W^e}(6) = \{1\}$, and $C_{W^e}(7) = \{4, 5\}$. Thus, $D_{W^d}(6,7) = \{1, 2, 3, 4, 5, 6, 7\}$.

Corollary 15. *For any two instances* $(x_{l_1}, y_{l_1}), (x_{l_2}, y_{l_2}) \in T$, *such that* $y_{l_1} \neq y_{l_2}$, *the row distance set* $D_{W^d}(l_1, l_2)$ *in the decoding matrix* W^d *equals the empty set, if* $C_{W^e}(l_1) \cap C_{W^e}(l_2) = \emptyset$ *and* $C_{W^e}(l_1) \cup C_{W^e}(l_2) = D_{W^e}(l_1, l_2)$.

Corollary 15 states that the row distance set $D_{W^d}(l_1, l_2)$ for instances (x_{l_1}, y_{l_1}), $(x_{l_2}, y_{l_2}) \in T$ is minimal (i.e., equal to $\emptyset$), if the two sets of the binary classifiers are disjointed and their union equals $D_{W^e}(l_1, l_2)$. Thus, if this condition holds for any two instances, the minimal row Hamming distance in the decoding matrix W^d equals 0.

Example 16. Take row 5 (class y_2) and row 8 (class y_4) in W^e and W^d in Fig. 1. Then $D_{W^e}(5,8) = \{2, 4, 5, 7\}$, $C_{W^e}(5) = \{4\}$, and $C_{W^e}(8) = \{2, 5, 7\}$. Thus, $D_{W^d}(5,8) = \emptyset$.

From Theorem 9 and Corollaries 11, 13, 15 given above we may conclude that for any two instances $(x_{l_1}, y_{l_1}), (x_{l_2}, y_{l_2}) \in T$, if $y_{l_1} \neq y_{l_2}$, we have that:

- the row distance set $D_{W^d}(l_1, l_2)$ (and the Hamming distance $|D_{W^d}(l_1, l_2)|$) does not depend directly on the size of the error sets $C_{W^e}(l_1)$ and $C_{W^e}(l_2)$ of the binary classifiers that err on these instances;
- the row distance set $D_{W^d}(l_1, l_2)$ (and the Hamming distance $|D_{W^d}(l_1, l_2)|$) depends on the sizes of the error-set differences $C_{W^e}(l_1) \setminus C_{W^e}(l_2)$ and $C_{W^e}(l_2) \setminus C_{W^e}(l_1)$;
- the row distance set $D_{W^d}(l_1, l_2)$ (and the Hamming distance $|D_{W^d}(l_1, l_2)|$) grows, when the error-set differences $C_{W^e}(l_1) \setminus C_{W^e}(l_2)$ and $C_{W^e}(l_2) \setminus C_{W^e}(l_1)$ get smaller for the set $D_{W^e}(l_1, l_2)$ and bigger for the set $D'_{W^e}(l_1, l_2)$ (Example 10).

To make a bigger picture: assume that the errors represented by the error sets $C_{W^e}(l_1)$ and $C_{W^e}(l_2)$ are more systematic; i.e., the binary classifiers $h_m \in C_{W^e}(l_1)$ are positively error correlated as well as the binary classifiers $h_m \in C_{W^e}(l_2)$. Then:

- the Hamming row distances in the decoding matrix W^d do not depend directly on the error correlation of the binary classifiers $h_m \in h$;
- the Hamming row distances in the decoding matrix W^d depend on the extent of overlap of sets of error-correlated binary classifiers $h_m \in h$;

– the Hamming row distances in the decoding matrix W^d grow, when the overlap of sets of error-correlated binary classifiers $h_m \in h$ increases.

Thus, our main conclusion is that the encoding matrix W^d of the *decoding* instance decomposition scheme $SP^d(T)$ can handle the error-correlated binary classifiers if they err in a similar way on instances of different classes. In this case, the instance decomposition schemes have error correcting capabilities and thus can be considered instance decomposition schemes of error-correcting output codes.

5   Experiments

To assess the generalization performance of the ensembles based on the instance-based decomposition schemes (IDS) we performed two sets of experiments. The first set of experiments in Sect. 5.1 studies how the error correlation of the binary classifiers influences the generalization performance of ensembles based on ECOC and ensembles based on IDS. The second set of experiments in Sect. 5.2 compares the generalization performance of the same ensembles for data with large number of classes.

Both sets of experiments have the following settings. The class decompositions schemes employed were exhaustive ECOC (eECOC) [5] and random ECOC (rECOC) [8]. The instance decomposition schemes (IDSs) employed were those that initialize instance encoding matrix based on either eECOC or rECOC (i.e., the instances for each class received initially the same code coming from eECOC/rECOC). eECOC were used for up to 9 classes and rECOC were used for more than 9 classes. Two types of classifiers were employed as binary base classifiers: the Ripper rule classifier[1] [3] and Logistic Regression [9]. The ensemble evaluation method was 10-fold cross validation averaged over 10 runs. The classification accuracy of the classifiers was compared using the paired t-test [12] at the 5 % significance level.

5.1   Error Correlation of Binary Classifiers vs. Ensemble Generalization Performance

The purpose of the set of experiments in this section is to study how the error correlation of the binary classifiers influences the generalization performance of the ensembles based on eECOC and IDSs. For the experiments IDSs were initialized using eECOC.

Our first experiment was on the Glass data from UCI [2]. The binary classifiers employed in the eECOC and IDS ensembles were Ripper classifiers. We recorded the accuracy of the ensembles while decreasing complexity of the Ripper binary classifiers. The complexity was controlled with the Ripper parameter minNumObj in the range $[0, 100]$[2]. We note that decreasing complexities of the

---

[1] Ripper was originally proposed as a binary classifier in [3].

[2] minNumObj is the minimal number of training instances to create a Ripper rule. Increasing the value of minNumObj decreases the Ripper complexity.

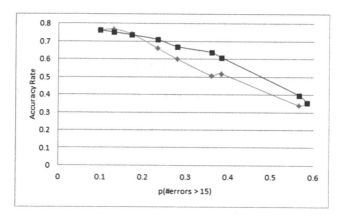

Fig. 2. The accuracy rate vs. the cumulative probability of $p(\#errors > 15)$ of the eECOC and IDS ensembles. The curve of the eECOC ensemble is denoted by ◆. The curve of the IDS ensemble is denoted by ■.

binary classifiers increases positive error-correlation levels between the classifiers. However, estimating the error correlation for more than two classifiers is difficult. Therefore, we employed instead the probability of the joint error of the binary classifiers, since it grows with that correlation. Thus, for each run of ensembles we recorded the empirical distribution of the joint-error probability of the binary Ripper classifiers.

To present the results, we note that the Glass data has 7 classes ($K = 7$). This implies that the eECOC ensemble can correct strictly at most 15 number of errors of the binary classifiers per test instance ($2^{K-3} - 1 = 2^{7-3} - 1$). That is why we plotted in Fig. 2 the accuracy of eECOC and IDS ensembles against the cumulative probability of $p(\#errors > 15)$. The figure shows that the accuracy of the IDS ensemble gets bigger than that of the eECOC ensemble for $p(\#errors > 15)$ greater than 0.1723. The IDS ensemble outperforms maximally the eECOC ensemble for $p(\#errors > 15)$ equal to 0.36 (the accuracy-rate difference was 0.13). Then the accuracy rates of the ensembles gradually converge and for $p(\#errors > 15)$ equal to 0.58 become equal. After this point the binary classifiers become majority-vote classifiers and the accuracy rates of the eECOC and IDS ensembles stay equal.

We observed similar behavior of the eECOC and IDS ensembles in our next experiments with 9 other data sets from UCI [2]. The results are in Tables 1 and 2.

Table 1 shows the accuracy rate of the ensembles in function of complexity of Ripper binary classifiers controlled by parameter minNumObj. The IDS ensembles won in 38 cases (18 times significantly) and lost in 6 cases. The number of draws was 16.

Table 2 shows the accuracy rate of the ensembles in function of complexity of Logistic-Regression binary classifiers controlled by the ridge parameter r in $[1, 50]$. The IDS ensembles won in 50 cases (14 times significantly) and lost in 7 cases. The number of draws was 3.

Table 1. The accuracy rate of the eECOC and IDS ensembles vs. complexity of Ripper binary classifiers controlled by parameter minNumObj. Bold numbers indicate statistically better results in group for a value of minNumObj.

Data	minNumObj=0		minNumObj=10		minNumObj=30		minNumObj=50		minNumObj=75		minNumObj=100	
	IDS	eECOC	IDS	eECOC	IDS	eECOC	IDS	eECOC	IDS	eECOC	IDS	eECOC
Anneal (7)	0,99	0,99	0,97	0,96	0,95	0,95	0,92	0,91	0,85	0,84	0,85	0,84
Autos (7)	0,84	0,84	0,79	0,76	**0,69**	0,56	0,54	0,48	**0,46**	0,38	0,33	0,33
Car (4)	0,91	0,9	**0,87**	0,83	**0,81**	0,78	0,78	0,77	0,79	0,77	**0,76**	0,74
Derm (6)	0,96	0,96	0,95	0,94	0,94	0,92	0,94	0,93	0,94	0,92	**0,95**	0,9
Ecoli (8)	0,84	0,84	0,81	0,84	0,81	0,81	**0,79**	0,75	**0,79**	0,69	**0,71**	0,65
Glass (7)	0,76	0,76	0,74	0,74	**0,67**	0,6	**0,61**	0,52	**0,4**	0,34	0,35	0,35
Iris (3)	0,91	0,91	0,94	0,94	0,92	0,92	0,33	0,33	0,33	0,33	0,33	0,33
Jap.Vow (9)	0,77	0,78	0,74	0,72	**0,69**	0,63	**0,67**	0,54	**0,58**	0,33	**0,34**	0,16
Lymph(4)	0,79	0,8	0,78	0,78	0,73	0,68	0,55	0,55	0,55	0,55	0,55	0,55
Zoo (7)	0,94	0,94	**0,96**	0,89	**0,88**	0,76	0,41	0,41	0,41	0,41	0,41	0,41

Analyzing the results we observe that the IDS ensembles outperform the eECOC ensembles especially when the complexity of the binary classifiers decreases. Decreasing the binary-classifier complexity increases the positive error correlation of the binary classifiers. Thus, the IDS ensembles can handle the error-correlated binary classifiers better than the eECOC ensembles in our experiments.

5.2 Ensemble Generalization Performance for Large Number of Classes

This section provides a set of experiments with ensembles based on ECOC and IDS for data with large number of classes. Due to the number of classes we employed random ECOC (rECOC) [8]. Hence, the initialization of the encoding

Table 2. The accuracy rate of the eECOC and IDS ensembles vs. complexity of Logistic-Regression binary classifiers controlled by the ridge parameter r in $[1, 50]$. Bold numbers indicate statistically better results in group for a r-value.

Data	r=1		r=10		r=20		r=30		r=40		r=50	
	IDS	eECOC	IDS	eECOC	IDS	eECOC	IDS	eECOC	IDS	eECOC	IDS	eECOC
Anneal (7)	0,99	0,99	0,97	0,97	0,96	0,95	**0,95**	0,93	**0,94**	0,92	**0,93**	0,9
Autos (7)	0,73	0,71	0,69	0,65	0,69	0,63	0,66	0,62	0,65	0,61	0,64	0,6
Car (4)	**0,94**	0,89	**0,92**	0,85	**0,9**	0,84	**0,88**	0,84	**0,88**	0,83	**0,87**	0,83
Derm (6)	0,98	0,975	0,98	0,98	0,98	0,98	0,98	0,98	0,97	0,98	0,9	0,97
Ecoli (8)	0,83	**0,87**	0,82	0,8	0,79	0,77	0,78	0,75	**0,78**	0,73	0,78	0,71
Glass (7)	0,59	0,62	0,58	0,58	0,58	0,57	0,58	0,57	0,6	0,56	0,61	0,56
Iris (3)	0,95	0,91	0,9	0,87	0,9	0,84	0,89	0,84	0,86	0,84	0,83	0,84
Jap.Vow (9)	0,83	0,8	0,79	0,75	0,75	0,72	0,74	0,69	0,73	0,67	**0,72**	0,65
Lymph(4)	0,84	0,86	0,84	0,84	0,84	0,84	0,84	0,83	0,84	0,82	0,83	0,82
Zoo (7)	0,9	0,9	0,9	0,88	0,9	0,88	**0,91**	0,84	**0,89**	0,82	**0,89**	0,8

Table 3. The accuracy rate of the eECOC and IDS ensembles vs. complexity of Logistic-Regression binary classifiers controlled by the ridge parameter r in $[1, 50]$. Bold numbers indicate statistically better results in group for a r-value.

Ripper	minNumObj=0		*Logistic regression*	minNumObj=0	
Data	IDS	eECOC	Data	IDS	eECOC
Abalone (28)	0.227	**0.261**	Abalone (28)	0,249	0,229
Letter (17)	**0.62**	0.559	Letter (17)	**0.61**	0,429
Patents (62)	0.235	0.226	Patents (62)	**0.246**	0,222
Pendigits (10)	0.86	0.81	Pendigits (10)	0.83	0,785

matrix of IDS was realized with rECOC. The binary classifiers Logistic Regression and Ripper were used with their default parameter settings: for Logistic Regression the ridge parameter r was set to 10 and for Ripper reduce-error pruning with 3 folds was set.

The results are given in Table 3 for 4 data sets with 10 to 62 classes. Among the 8 experiments the IDS ensembles won in 7 cases (3 times significantly). These results are expected, since for large number of classes the probability of having difficult binary classification tasks increases. Thus, the experiments confirm that the IDS ensembles can handle better difficult binary classification problems.

6   Conclusion

In this paper we proposed instance decomposition schemes (IDSs) for mapping multi-class classification tasks into a series of binary classification tasks. We showed theoretically and experimentally that IDSs are capable of handling the two main problems of the class decomposition schemes: the problem of difficult binary classification tasks and the problem of positive error correlation of the binary classifiers.

Future research will focus on computational efficiency of IDS. We note that IDS employs nearest neighbor classification (based on Hamming distance). Therefore, two research directions are foreseen: (1) to reduce the number of rows and (2) to reduce the number of columns in the decoding matrices. Several techniques are readily available from edited nearest neighbor rules [11] and multi-variate feature selection [7].

References

1. Alpaydin, E., Mayoraz, E.: Learning error-correcting output codes from data. In: Proceedings of the Ninth International Conference on Artificial Neural Networks (ICANN-99), pp. 743–748. MIT Press (1999)
2. Bache, K., Lichman, M.: UCI machine learning repository (2013)
3. Cohen, W.W.: Fast effective rule induction. In: Proceeding of the Twelfth International Conference on Machine Learning, pp. 115–123. Morgan Kaufmann (1995)

4. Dekel, O., Singer, Y.: Multiclass learning by probabilistic embeddings. Adv. Neural Inf. Process. Syst. **15**, 945–952 (2002)
5. Dietterich, T.G., Bakiri, G.: Solving multiclass learning problems via error-correcting output codes. J. Artif. Intell. Res. **2**, 263–286 (1995)
6. Escalera, S., Pujol, O., Radeva, P.: Error-correcting ouput codes library. J. Mach. Learn. Res. **11**, 661–664 (2010)
7. Guyon, I., Elisseeff, A.: An introduction to variable and feature selection. J. Mach. Learn. Res. **3**, 1157–1182 (2003)
8. Hall, M., Frank, E., Holmes, G., Pfahringer, B., Reutemann, P., Witten, I.H.: The WEKA data mining software: an update. SIGKDD Explor. **11**, 10–18 (2009)
9. le Cessie, S., van Houwelingen, J.C.: Ridge estimators in logistic regression. Appl. Stat. **41**(1), 191–201 (1992)
10. Lorena, A.C., de Carvalho, A., Gama, J.M.P.: A review on the combination of binary classifiers in multiclass problems. Artif. Intell. Rev. **30**, 19–37 (2008)
11. Marchiori, E.: Hit miss networks with applications to instance selection. J. Mach. Learn. Res. **9**, 997–1017 (2008)
12. Nadeau, C., Bengio, Y.: Inference for the generalization error. In: Solla, S.S., Leen, T.K., Müller, K.-R. (eds.) Advances in Neural Information Processing Systems 12, pp. 307–313. The MIT Press, Cambridge (1999)
13. Pujol, O., Radeva, P., Vitrià, J.: Discriminant ECOC: a heuristic method for application dependent design of error correcting output codes. IEEE Trans. Pattern Anal. Mach. Intell. **28**(6), 1007–1012 (2006)
14. Rätsch, G., Smola, A.J., Mika, S.: Adapting codes and embeddings for polychotomies. Adv. Neural Inf. Process. Syst. **15**, 513–520 (2002)
15. Zhou, J., Peng, H., Suen, C.Y.: Data-driven decomposition for multi-class classification. Pattern Recogn. **41**(1), 67–76 (2008)
16. Zor, C., Yanikoglu, B.A., Windeatt, T., Alpaydin, E.: FLIP-ECOC: a greedy optimization of the ECOC matrix. In: Proceedings of the 25th International Symposium on Computer and Information Sciences, London, UK, 22–24 September 2010, pp. 149–154 (2010)

Pruning Bagging Ensembles with Metalearning

Fábio Pinto[(⊠)], Carlos Soares, and João Mendes-Moreira

INESC TEC/Faculdade de Engenharia, Universidade Do Porto,
Rua Dr. Roberto Frias, S/n, 4200-465 Porto, Portugal
fhpinto@inesctec.pt, {csoares,jmoreira}@fe.up.pt

Abstract. Ensemble learning algorithms often benefit from pruning strategies that allow to reduce the number of individuals models and improve performance. In this paper, we propose a Metalearning method for pruning bagging ensembles. Our proposal differs from other pruning strategies in the sense that allows to prune the ensemble before actually generating the individual models. The method consists in generating a set characteristics from the bootstrap samples and relate them with the impact of the predictive models in multiple tested combinations. We executed experiments with bagged ensembles of 20 and 100 decision trees for 53 UCI classification datasets. Results show that our method is competitive with a state-of-the-art pruning technique and bagging, while using only 25 % of the models.

Keywords: Ensemble learning · Metalearning · Classification · Pruning

1 Introduction

Ensemble learning (EL) refers to methods that combine several models to make a final prediction, typically in a classification or regression scenario. The EL literature can be split into three main topics: ensemble generation, ensemble pruning and ensemble integration. This paper proposes a Metalearning (MtL) method to prune bagging ensembles of classifiers. Our approach differs from the other ensemble pruning methods in the sense that allows to prune the ensemble by just analyzing the characteristics of a bootstrap sample and before actually generating the individual models.

Combining complementary classifiers can improve the accuracy over individual models. One can say that two classifiers are complementary if they make errors in different regions of the input space [1]. For complementarity between classifiers there is a need for diversity. Several measures have been proposed in the literature to quantify the concept of diversity [2]. Research on this topic has inspired the development of methods for ensemble pruning. The aim of these methods is to find a subset of models that improves or at least has the same generalization ability of the full set of models. Several techniques have been proposed in this scope [3,4], particularly for bagging [5].

Bagging is an EL technique that allows to generate multiple predictive models and aggregate their output to provide a final prediction [6]. Typically, the

© Springer International Publishing Switzerland 2015
F. Schwenker et al. (Eds.): MCS 2015, LNCS 9132, pp. 64–75, 2015.
DOI: 10.1007/978-3-319-20248-8_6

aggregation function is the mean (if the outcome is a quantitative variable) or the mode (if the outcome is a qualitative variable). The models are built by applying a learning algorithm to bootstrap replicates of the learning set. Diversity is achieved through the use of different bootstrap samples.

We developed a MtL method that predicts if a given bootstrap sample of a dataset is going to originate an useful classifier. By useful, in the scope of this paper, should be understood as a classifier that is accurate in a specific region of the input space. For that, we compute a set of bootstrap characteristics together with a variable that represents the usefulness of those bootstraps. This variable is computed by carrying exhaustive experiments in which we test several combinations of bagged ensembles of decision trees with 20 and 100 models. We used 53 UCI [7] classification datasets for our experiments. The method is evaluated on the same 53 datasets using a leave-one-out methodology.

The main contributions of this paper are: *(1)* a MtL method for pruning bagging ensembles *(2)* metafeatures specifically developed for MtL in the context of ensemble problems *(3)* comparison of different methods for ensemble pruning in 53 classification datasets.

The paper is organized as follows. In Sect. 2 we present an overview of the literature both on ensemble pruning and MtL. In Sect. 3 we describe the MtL method for pruning bagging ensembles. Section 4 presents our methodology. In Sect. 5 we show the experimental results and discussion. Finally, in Sect. 6, we conclude the paper and define future work.

2   Related Work

For classification, ensemble pruning methods are often inspired by the ensemble learning diversity literature. That is, the methods focus on searching for complementary classifiers. Margineantu and Dietterich [8] showed firstly that there is no need for all the classifiers in a boosting ensemble. The development of ensemble pruning methods are often biased towards bagging since it is noted that these kind of methods are more effective with bagging than with boosting [9].

In terms of the nature of the methods, we can see two different research directions: one focused on optimization based methods and another on ordering based methods. In the former, the methods use an optimization technique to select a subset of models. Zhang et al. [4] approached ensemble pruning as a quadratic integer programming problem that is solved by applying semi-definite programming to a convex relaxation of the original problem. Chao et al. [10] proposed a method inspired by a multi-objective evolutionary algorithm. In the latter, the methods start with an empty and gradually add models in order to minimize/maximize a certain objective. This iterative process allows to generate an order of the individual models. Martinez-Muñoz [5] published a detailed analysis of these kind of methods for bagging ensembles. We use one of the methods (the one with the best overall results in [5] - *MDSQ*) proposed by them for comparison with our method. Li et al. [11] present a theoretical study of diversity for ensembles of classifiers and proposed a greedy forward pruning method that exploits their discoveries.

Another important question regarding ensemble pruning that is relevant for most of the methods is the size of the pruned ensemble. Hernández-Lobato et al. [12] empirically showed that the optimal ensemble size is very sensitive to the particular classification problem considered. However, for a wide range of classification problems, a pruning of 60–80% seems appropriate [5].

Although the use of ensembles has been becoming more popular due to their superior performance, it is still possible to improve the generalization ability of an ensemble by pruning or use a specific integration strategy [13]. One of the fields that can assist this process is MtL. As the study of principled methods that exploit metaknowledge to obtain efficient models and solutions by adapting machine learning and data mining processes [14], MtL can help in developing useful methods for ensemble learning while still providing interesting and useful knowledge about the ensemble and the problem domain.

The most widely known application of MtL to ensemble learning are the Meta-Decision Trees (MDT), proposed by Todorovski and Džeroski [15]. They presented an algorithm for learning a decision tree based on C4.5 that instead of making a prediction, the leaves of the tree specify which classifier should be used to obtain a prediction. Their study comprised 21 classification datasets and 5 base-level classifiers. Results show that MDT are better than voting and stacking in combining classifiers, while still providing comprehensible knowledge about the ensemble and the predictive problem.

The main issue in MtL is defining the metafeatures. The most used ones are the simple, statistical and information-theoretic metafeatures [14]. This set includes the *number of attributes* of the dataset, *mutual information between symbolic attributes* or *class entropy*, to name a few. This kind of metafeatures has the advantage of providing interpretable knowledge about the problems. Another kind of metafeatures are the model-based ones [16]. These capture some characteristic of a model generated by applying a learning algorithm to a dataset, *i.e., the number of leaf nodes of a decision tree*. Finally, a metafeature can also be a landmarker [17]. These are generated by making a quick performance estimate of a learning algorithm in a particular dataset. In this paper we introduce new metafeatures that are specific to our problem.

3   A Metalearning Method for Ensemble Pruning

Formally, an ensemble F gathers a set of predictors of a function f denoted as $\hat{f}_i$. Therefore,

$$F = \{\hat{f}_i, i = 1, ..., k\}$$

where the ensemble predictor is defined as $\hat{f}_f$.

In [18], we propose a methodology to empirically analyze the behavior of bagging. Given a set of k bootstrap samples (also referred to in this paper as bootstraps, for simplicity), we estimate the empirical distribution of performance of the bagging ensembles that can be generated from all elements of its power set. In other words, we estimate the empirical distribution of performance of all possible ensembles of size 2, 3, ..., k that can be generated from those k bootstraps.

This distribution can be used to study the role of a given bootstrap (and respective predictive model $\hat{f}_i$) in the performance of $2^k - 1$ possible ensembles, as done in this paper. Additionally, the distribution can be used to analyze the joint relationship between the bootstrap samples in each ensemble and its performance.

We sampled (stratified) all the 53 UCI datasets with more than 10000 instances for computational reasons. The error estimation methodology was hold-out[1]: 60 % for training set, 20 % for validation set and 20 % for the test set. Each ensemble is evaluated in the test set using accuracy as error measure. We executed experiments in which we tested all possible combinations of ensembles with $k = 20$ and sampled the $k = 100$ case. It is easy to understand that is impossible to execute the complete set of experiments for ensembles with a realistically large size, such as $k = 100$, given that the number of combinations to test is $2^k - 1$. Therefore, the only possibility is to estimate the distribution of the performance of all ensembles that can be generated with the set of k bootstraps by sampling from its power set. We provide a study of the effectiveness of this sampling procedure in [18]. In this paper, we execute experiments for both cases ($k = 20$ and $k = 100$) and we compare results.

3.1 Metatarget

The experiments that we carried with the UCI datasets allowed to collect the distribution of the performance of the bagging algorithm in very distinct learning problems. Our goal is to quantify the importance of each model (and respective bootstrap) in the ensemble space. Then, we need to aggregate the results obtained for each one of them and compute an estimate of importance.

We adapted the measure NDCG [19] (Normalized Discounted Cumulative Gain) to form our metatarget. We consider the performance of the ensembles (in decreasing order) to which the bootstrap k belongs, for each dataset, as $acc_{1,d}$, $acc_{1,d}$, ..., $acc_{n,d}$ where n represents an ensemble and d a dataset. Therefore, for each bootstrap k of the dataset d, we calculate the respective DCG

$$DCG_{k,d} = \sum_{n=1}^{100} acc_{n,d} + \sum_{101}^{n} \frac{acc_{n,d}}{log_{100}n}$$

and we normalize it by an ideal ranking ($IDCG_d$) in which the best ensembles for each dataset are selected. Then,

$$NDCG_{k,d} = \frac{DCG_{k,d}}{IDCG_d}.$$

3.2 Metafeatures

For this work, we relied on simple, statistical, information-theoretic and landmarker metafeatures. For the first group, we selected several metafeatures

---

[1] For computational reasons, it was possible to apply a cross-validation methodology.

already present in the literature which were first used for MtL in the METAL and Statlog projects [14]. We also introduce a new metafeature based on the Jensen-Shannon distance [20] between a bootstrap and the training set. This metafeature aims to measure how different is the bootstrap from the original dataset. This metric has proved to be useful while measuring stability in multi-source data [21]. It can also be seen as a diversity measure of low order that focuses directly on the bootstrap sample and not on the predictions made by the generated model.

However, we also used other types of metafeatures. Two landmarkers: a decision stump and a Naive Bayes classifier. Given the different bias of the algorithms, it is expected that the metafeatures can help capture different patterns [22]. We also used two diversity measures proposed in the ensemble learning literature: the Q-Statistic [2] and Classifier Output Difference [23] (COD) measures. Kuncheva et al. [2] state that the Q-Statistic is the diversity measure with greater potential for providing useful information about ensemble performance. However, in the same paper, the authors claim that the usefulness of diversity measures in building classifier ensembles for real-life pattern recognition problems is questionable. We adapted the Q-Statistic to the specificities of our problem. Kuncheva et al. present it as a metric to measure the diversity of an ensemble. We use it to measure the diversity between the predictions of two landmarker models (Naive Bayes): one generated by applying a learning algorithm to a bootstrap and the other to the original dataset. Using such measure in this study gives a different perspective on its usefulness. Formally, our adapted version of the Q-Statistic is defined as

$$Q_{b,d} = \frac{N^{bb}N^{dd} - N^{db}N^{bd}}{N^{bb}N^{dd} + N^{db}N^{bd}}$$

where each element is formed as in Table 1.

Table 1. Relationship between a pair of classifiers.

	$f_b$correct	$f_d$correct
$f_b$correct	N^{bb}	N^{bd}
$f_d$correct	N^{db}	N^{dd}

The COD metric has been proposed as a measure to estimate the potential of combining classifiers:

$$\hat{COD}_T(\hat{f}_b, \hat{f}_d) = \frac{\sum_{x \in Ts} \begin{cases} 1, & \text{if } \hat{f}_b(x) = \hat{f}_d(x) \\ 0, & \text{otherwise} \end{cases}}{|Ts|}$$

Lee and Giraud-Carrier [24] published a paper on unsupervised MtL in which they study the application of several diversity measures for ensemble learning as a distance function for clustering learning algorithms. In their experiments, only one measure, COD, presents results that indicate that it can be a good measure for this kind of task. This is indicative that the metric can also be useful in our problem.

Therefore, the set of metafeatures used for this work is:

- number of examples
- number of attributes
- proportion of symbolic attributes
- proportion of missing values
- proportion of numeric attributes with outliers
- class entropy
- average entropy between symbolic attributes
- average mutual information between symbolic attributes and the class
- average mutual information between pairs of symbolic attributes
- average absolute correlation between numeric attributes
- average absolute skewness between numeric attributes
- average kurtosis between numeric attributes
- canonical correlation of the most discriminating single linear combination of numeric attributes and the class distribution
- Jensen-Shannon distance between the dataset and bootstrap
- Decision Stump landmarker
- Naive Bayes landmarker
- Q-Statistic diversity
- COD diversity

4   Methodology

This Section specifies the methodology used to evaluate our pruning method. The problem is addressed as a regression task at the meta-level and as a classification task at the base-level.

4.1   Error Estimation

The system was evaluated with leave-one-out cross-validation. However, in each fold, instead of leaving a single instance for testing, all the instances associated with the same dataset are used for testing. This procedure allows to train the meta-model in meta-data from 52 datasets and test the approach in another dataset, iteratively. The final error estimation is computed by averaging the results on the 53 datasets.

At the meta-level, since the target is a numeric variable, the error measure is the *Root Mean Squared Error* (*RMSE*), defined as

$$\sqrt{\frac{\sum_{j=1}^{m}(y_j - \hat{y}_j)^2}{m}}$$

where y_j is the true value, $\hat{y}_j$ is the predicted value and m the number of testing meta-instances.

At the base-level, the error measure is accuracy, defined as

$$accuracy = \frac{TP + TN}{TP + FP + TN + FN}$$

where TP are the true positives, TN the true negatives, FP the false positives and FN the false negatives. Both evaluation procedures (meta-level and base-level) are going to be carried with the assistance of the methodology proposed by Demšar [25].

4.2 Meta-Learners

We used three learning algorithms to generate the meta-models: M5' [26] (*Meta.M5'*), Support Vector Machines with radial basis kernel function [27] (*Meta.SVM*) and Random Forests [28] (*Meta.RF*). The performance of the meta-learners is going to be compared with a baseline: the average of the metatarget in the (meta) training data.

4.3 Benchmark Pruning Methods

We compare our method with 4 alternatives:

- *Metatarget.* In this approach we use the groundtruth of our metatarget to execute the pruning at the basel-level. This allows to benchmark how good our method could be if we were able to generate an idealistic meta-model.
- *Bagging.* The same algorithm proposed by Breiman [6], without any sort of pruning.
- *Margin Distance Minimization (MDSQ)* [5]. This algorithm belongs to the family of pruning methods base on modifying the order in which classifiers are aggregated in a bagging ensemble. The main feature of these kind of methods is to exploit the complementariness of the individual classifiers and find a subset with good performance. Results presented by the authors show that *MDSQ* can greatly reduce the size of the ensemble without significant loss of generalization ability (for some datasets an improvement of the results over bagging was verified).
- *Random Pruning.* A baseline approach in which the selection of models to be pruned is random. This is repeated 30 times for robust results.

4.4 Ensemble Size

As mentioned previously, determining the optimal ensemble size is not a trivial task. Hernández-Lobato et al. [12] show that the optimal size of a bagging ensemble for a dataset is very specific to the particular classification problem considered. However, taking into account their results in 25 datasets, we concluded that a pruning percentage of 75 % of the ensemble should allow a good performance of all the methods in the majority of the classification problems. So, all the pruning methods tested in this paper follow this rule.

5   Experiments

All the experiments were carried in the R software [29], using the package *party* for generating the decision trees. We used RReliefF [30] to assist the selection of metafeatures both for $k = 20$ and $k = 100$.

5.1   Meta-Level Results

As mentioned previously, we compared the performance of the meta-model with a baseline using RMSE as error measure. To assess if the meta-model is significantly better than the baseline, we used the methodology proposed by Demšar [25] with $\alpha = 0.05$.

Figure 1 shows the Critical Difference (CD) diagrams of the results obtained at the meta-level. For $k = 20$, we can see that both M5' and RF present a better performance that SVM and the baseline. The difference in terms of performance between M5' and RF is not statistically significant. The same can be stated for SVM and the baseline. For $k = 100$, the same result can be verified, although in this case RF shows a slightly better performance than M5'. Again, yet, this difference is not statistically significant.

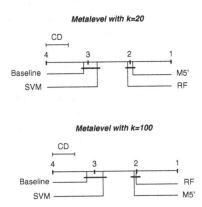

Fig. 1. Critical Difference diagrams of the performance of the meta-models in comparison with the baseline, at the meta-level.

This result shows that the metafeatures that we generated for this problem are informative and can possibly be used to predict the usefulness of a model generated from a bootstrap sample.

5.2   Base-Level Results

The base-level evaluation of the method was, again, carried with the methodology proposed by Demšar [25] with $\alpha = 0.05$.

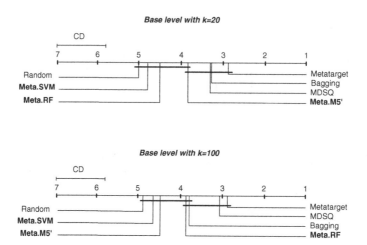

Fig. 2. Critical Difference diagrams of the performance of the metamodels in comparison with the benchmark pruning methods, at the base-level.

Figure 2 shows the CD diagrams for the base-level results. For $k = 20$, our method achieves the best performance with the M5' learning algorithm (this is in agreement with the results obtained at the meta-level for $k = 20$). The performance of *Meta.M5'* is worst than *Metatarget*, *Bagging* and *MDSQ*, although this difference is not statistically significant. Comparing with the *Random* benchmark, *Meta.M5'* shows better performance. However, the difference between the generalization ability of the method is not statistically significant.

For $k = 100$, our method shows better performance while pairing with the RF learning algorithm (again, this is in agreement with the results obtained at the meta-level for $k = 20$). The results are very similar with the ones obtained in the $k = 20$ scenario. The performance of *Meta.RF* is worst than *Metatarget*, *MDSQ* and *Bagging*, although this difference is not statistically significant. Regarding the comparison with the *Random* baseline, *Meta.RF* presents better performance but the difference is not statistically significant.

The performance of *Bagging* and *MDSQ* is very similar both for $k = 20$ and $k = 100$. *MDSQ* shows a better performance in the $k = 100$ scenario. This result is expected because the method needs a reasonable number of models in order to achieve good performance [5]. This fact also increases the computational cost of the method.

The pruning executed with the *Metatarget* shows the best performance for both cases, $k = 20$ and $k = 100$. This is indicative that our method, with adequate metafeatures, can become a very useful and innovative pruning technique.

5.3   Discussion

Our work is the first approach using MtL that attempts to understand the behavior of ensemble learning algorithms and improve them.

The results presented state that our method already has the ability to prune bagged ensembles of decision trees with a performance competitive with the state-of-the-art algorithms. However, there is room for improvement since its performance is not far superior from a naive baseline such as *Random*. Furthermore, since the *Metatarget* benchmark surpasses all the methods tested, it is our goal to further improve the performance of our method in that direction.

We believe that one of the key aspects that affect the most our results is the fact that we are trying to predict the usefulness of one model instead of a subset. In the latter scenario, we could use diversity measures already proposed in the ensemble learning literature as metafeatures and take into account the complementarity between more than two classifiers. The diversity metafeatures that we use in this paper such as *COD* or *Q-Statistic* only relate the landmarker models generated from the bootstrap samples with the landmarker models generated from the original training data. We plan to investigate an approach in which our method predicts the accuracy of subsets of models instead of the usefulness of individual ones as we present here.

6   Conclusions

This paper proposes a MtL method for pruning bagging ensembles of decision trees. Our proposal differs from the other methods proposed in the literature in the sense that allows to prune the ensemble before actually generating the predictive models. This feature can be particularly important in contexts with limited computational resources, such as online applications [31].

We tested our method against bagging and a state-of-the-art pruning technique, *MDSQ*. Results show that our method is competitive with bagging (using only 25 % of the bagged models) and *MDSQ* (with less computational cost since it does not require to generate all the models of the bagged ensemble). Furthermore, as a *topline* approach, we tested the performance of the method using the groundtruth of the metatarget. The evaluation of all the methods showed that if we can generate more informative metafeatures for this problem, the MtL method can surpass *MDSQ* while still being computationally less demanding. This is our main future work.

As mentioned before, we plan to adapt the method so that instead of it predicting the usefulness of individuals models, it predicts the accuracy of subsets of models. We believe that this can enhance our results since it can greatly benefit from ensemble learning literature regarding diversity measures.

We also plan to extend and adapt the methodology proposed in this paper to other ensemble learning algorithms like boosting or random forests. This would bring challenges in the development of the metafeatures in order to deal with probabilistic processes like the ones that occur in boosting.

Finally, the methodology presented in this paper could be adapted for dynamic integration of ensembles. The metaknowledge extracted from an ensemble could be used for a more effective predictive performance while providing interpretability information about the domain.

Acknowledgements. This work is partially funded by FCT/MEC through PID-DAC and ERDF/ON2 within project NORTE-07-0124-FEDER-000059, a project financed by the North Portugal Regional Operational Programme (ON.2 O Novo Norte), under the National Strategic Reference Framework (NSRF), through the European Regional Development Fund (ERDF), and by national funds, through the Portuguese funding agency, Fundação para a Ciência e a Tecnologia (FCT) within project UID/EEA/50014/2013.

References

1. Brown, G., Wyatt, J., Harris, R., Yao, X.: Diversity creation methods: a survey and categorisation. Inf. Fusion **6**(1), 5–20 (2005)
2. Kuncheva, L.I., Whitaker, C.J.: Measures of diversity in classifier ensembles and their relationship with the ensemble accuracy. Mach. Learn. **51**(2), 181–207 (2003)
3. Zhou, Z.H., Tang, W.: Selective ensemble of decision trees. In: Wang, G., Liu, Q., Yao, Y., Skowron, A. (eds.) RSFDGrC 2003. LNCS (LNAI), vol. 2639, pp. 476–483. Springer, Heidelberg (2003)
4. Zhang, Y., Burer, S., Street, W.N.: Ensemble pruning via semi-definite programming. J. Mach. Learn. Res. **7**, 1315–1338 (2006)
5. Martinez-Muñoz, G., Hernández-Lobato, D., Suárez, A.: An analysis of ensemble pruning techniques based on ordered aggregation. IEEE Trans. Pattern Anal. Mach. Intell. **31**(2), 245–259 (2009)
6. Breiman, L.: Bagging predictors. Mach. Learn. **24**(2), 123–140 (1996)
7. Blake, C., Merz, C.J.: UCI repository of machine learning databases (1998)
8. Margineantu, D.D., Dietterich, T.G.: Pruning adaptive boosting. In: ICML, vol. 97, pp. 211–218. Citeseer (1997)
9. Hernández-Lobato, D., Hernández-Lobato, J.M., Ruiz-Torrubiano, R., Valle, Á.: Pruning adaptive boosting ensembles by means of a genetic algorithm. In: Corchado, E., Yin, H., Botti, V., Fyfe, C. (eds.) IDEAL 2006. LNCS, vol. 4224, pp. 322–329. Springer, Heidelberg (2006)
10. Qian, C., Yu, Y., Zhou, Z.H.: Pareto ensemble pruning. In: AAAI Conference on Artificial Intelligence (2015)
11. Li, N., Yu, Y., Zhou, Z.-H.: Diversity regularized ensemble pruning. In: Flach, P.A., De Bie, T., Cristianini, N. (eds.) ECML PKDD 2012, Part I. LNCS, vol. 7523, pp. 330–345. Springer, Heidelberg (2012)
12. Hernández-Lobato, D., Martínez-Muñoz, G., Suárez, A.: How large should ensembles of classifiers be? Pattern Recogn. **46**(5), 1323–1336 (2013)
13. Mendes-Moreira, J., Soares, C., Jorge, A.M., Sousa, J.F.D.: Ensemble approaches for regression: a survey. ACM Comput. Surv. (CSUR) **45**(1), 1–40 (2012). Article No. 10
14. Brazdil, P., Carrier, C.G., Soares, C., Vilalta, R.: Metalearning: Applications to Data Mining. Springer, Heidelberg (2008)
15. Todorovski, L., Džeroski, S.: Combining classifiers with meta decision trees. Mach. Learn. **50**(3), 223–249 (2003)
16. Peng, Y., Flach, P.A., Soares, C., Brazdil, P.: Improved Dataset Characterisation for Meta-learning. In: Lange, S., Satoh, K., Smith, C.H. (eds.) DS 2002. LNCS, vol. 2534, pp. 141–152. Springer, Heidelberg (2002)
17. Pfahringer, B., Bensusan, H., Giraud-Carrier, C.: Tell me who can learn you and i can tell you who you are: landmarking various learning algorithms. In: Proceedings of the 17th International Conference on Machine Learning, pp. 743–750 (2000)

18. Pinto, F., Soares, C., Mendes-Moreira, J.: An empirical methodology to analyze the behavior of bagging. In: Luo, X., Yu, J.X., Li, Z. (eds.) ADMA 2014. LNCS, vol. 8933, pp. 199–212. Springer, Heidelberg (2014)

19. Järvelin, K., Kekäläinen, J.: Cumulated gain-based evaluation of IR techniques. ACM Trans. Inf. Syst. (TOIS) **20**(4), 422–446 (2002)

20. Lin, J.: Divergence measures based on the shannon entropy. IEEE Trans. Inf. Theory **37**(1), 145–151 (1991)

21. Saez, C., Robles, M., Garcia-Gomez, J.M.: Comparative study of probability distribution distances to define a metric for the stability of multi-source biomedical research data. In: EMBC, pp. 3226–3229. IEEE (2013)

22. Fürnkranz, J., Petrak, J.: An evaluation of landmarking variants. In: ECML/PKDD 2000 Workshop on Integrating Aspects of Data Mining, Decision Support and Meta-Learning, pp. 57–68 (2001)

23. Peterson, A.H., Martinez, T.: Estimating the potential for combining learning models. In: Proceedings of the ICML Workshop on Meta-Learning, pp. 68–75 (2005)

24. Lee, J.W., Giraud-Carrier, C.: A metric for unsupervised metalearning. Intell. Data Anal. **15**(6), 827–841 (2011)

25. Demšar, J.: Statistical comparisons of classifiers over multiple data sets. J. Mach. Learn. Res. **7**, 1–30 (2006)

26. Wang, Y., Witten, I.H.: Inducing model trees for continuous classes. In: Proceedings of the Ninth European Conference on Machine Learning, pp. 128–137 (1997)

27. Cortes, C., Vapnik, V.: Support-vector networks. Mach. Learn. **20**(3), 273–297 (1995)

28. Breiman, L.: Random forests. Mach. Learn. **45**(1), 5–32 (2001)

29. R Core Team, : R: A Language and Environment for Statistical Computing. R Foundation for Statistical Computing, Vienna (2012). ISBN 3-900051-07-0

30. Robnik-Šikonja, M., Kononenko, I.: Theoretical and empirical analysis of ReliefF and RReliefF. Mach. Learn. **53**(1–2), 23–69 (2003)

31. Prodromidis, A.L., Stolfo, S.J.: Cost complexity-based pruning of ensemble classifiers. Knowl. Inf. Syst. **3**(4), 449–469 (2001)

Multi-label Selective Ensemble

Nan Li[✉], Yuan Jiang, and Zhi-Hua Zhou

National Key Laboratory for Novel Software Technology, Nanjing University
Collaborative Innovation Center of Novel Software Technology and Industrialization,
Nanjing 210023, China
{lin,jiangy,zhouzh}@lamda.nju.edu.cn

Abstract. Multi-label selective ensemble deals with the problem of
reducing the size of multi-label ensembles whilst keeping or improving
the performance. In practice, it is of important value, since the generated ensembles are usually unnecessarily large, which leads to extra high
computational and storage cost. However, it is more challenging than traditional selective ensemble, because real-world applications often employ
different performance measures to evaluate the quality of multi-label predictions, depending on user requirements. In this paper, we propose the
MUSE approach to tackle this problem. Specifically, by directly considering the concerned performance measure, we develop a convex optimization formulation and provide an efficient stochastic optimization solution
for a large variety of multi-label performance measures. Experiments show
that MUSE is able to obtain smaller multi-label ensembles, whilst achieving better or at least comparable performance in terms of the concerned
performance measure.

Keywords: Multi-label classification · Ensemble pruning · Selective
ensemble

1 Introduction

Multi-label learning deals with the problem where each instance is associated
with multiple labels simultaneously, and it has wide applications in different
domains, for example, document categorization where a document may belong
to multiple topics [16,27], multi-media annotation where an image or a music
can be annotated with more than one tags [1,26]. During the past few years,
it has become an active research topic [2,7,8,10,11,19,20,28,31], and a recent
comprehensive survey can be found in [32].

In multi-label learning, label correlations have been widely accepted to be
important [3,30], and many approaches have been proposed to exploit label
correlations. Amongst them, multi-label ensemble methods which construct a
group of multi-label classifiers and combine them for prediction have drawn
much attention. For examples, random k-labelsets (RAKEL) is an ensemble of
classifiers, each taking a small random subset of labels and learning on the power
set of this subset [25]; ensemble of pruned sets (EPS) combines a set of pruned

© Springer International Publishing Switzerland 2015
F. Schwenker et al. (Eds.): MCS 2015, LNCS 9132, pp. 76–88, 2015.
DOI: 10.1007/978-3-319-20248-8_7

	Component Classifiers
Hamming	□ ■ ■ □ ■ ■ □ □ ■ ■ □ ■ □ □ □ □ □ □ ■ ■
One-error	□ ■ ■ □ □ □ ■ □ ■ □ □ ■ ■ □ □ ■ □ ■ □ ■

Fig. 1. Comparison of the selected classifiers when considering different multi-label performance measures (i.e., Hamming loss and one-error), where the component classifiers are classifier chains with random label ordering trained on the *cal500* dataset, the selective ensemble chooses 9 out of 20 component classifiers via exhaustive search, and '■'/'□' indicates it is selected/unselected.

sets classifiers, each mapping sets of labels to single labels while pruning infrequent ones [18]; ensemble of classifier chains (ECC) combines multiple classifier chain classifiers, each learning an extended binary relevance classifier based on a random label ordering [19], and EPCC is a probabilistic extension of ECC [3]. From the perspective of ensemble learning, these multi-label ensemble methods try to construct diverse multi-label classifiers [22], mostly by smart heuristic randomization strategies.

In general, by combining more diverse multi-label classifiers, the ensemble performance tends to improve and converge. However, one issue is that the constructed ensembles tend to be unnecessarily large, requiring large amount of memory and also decreasing the response time of prediction. In traditional single label learning, selective ensemble (*a.k.a.* ensemble pruning or ensemble selection) addresses this issue by choosing a subset of component classifiers to form a subensemble, this has achieved success and became an active research topic in ensemble learning (see [34, chapter 6]). In this paper, we study the selective ensemble problem in the multi-label learning setting, that is, we try to reduce the size of a multi-label ensemble, such that compared with the original ensemble, the memory requirement and the response time can be reduced while similar or better prediction performance can be achieved.

In contrast to traditional supervised learning which usually takes accuracy as the performance measure, multi-label learning systems often employ different performance measures to evaluate the quality of multi-label predictions, depending on the application and user requirements [20]. These measures are designed from different aspects, and a classifier performing well in terms of one measure does not necessarily achieve good performance in terms of other measures, and it has been shown that a multi-label classifier tailored for one specific performance measure can perform poorly in terms of other measures [4]. This will make multi-label selective ensemble quite different from traditional selective ensemble, that is, we need to take the concerned performance measure fully into account when generating multi-label selective ensemble. As an example, we can see in Fig. 1 that the classifiers selected for Hamming loss are quite different from those for one-error. Essentially, this makes the task of multi-label selective ensemble more challenging, because most multi-label performance measures are quite complicated and difficult to optimize, for example, most of them are non-decomposable over labels, non-convex, and non-smooth. To deal with this issue, we propose the

MUSE approach to optimize the concerned performance measure. Specifically, by focusing on two groups of multi-label performance measures, we formulate the problem into a convex optimization problem with ℓ_1-norm regularization, and then present an efficient stochastic optimization solution. Experiments on real-world datasets show the effectiveness of the proposed MUSE approach.

The remainder of the paper is organized as follows. Section 2 presents our proposed MUSE approach. Section 3 reports the experiment results. Section 4 makes some brief discussion with related work, which is followed by the conclusion in Sect. 5.

2   The MUSE Approach

Let $\mathcal{X}$ be the instance space and $\mathcal{L}$ be a set of l labels. In multi-label learning, each instance $\mathbf{x}_i \in \mathcal{X}$ is associated with multiple labels in $\mathcal{L}$, which is represented as an l-dimensional binary vector $\mathbf{y}_i$ with the k-th element 1 indicating $\mathbf{x}_i$ is associated with the k-th label and -1 otherwise. Given a set of training examples $S = \{(\mathbf{x}_i, \mathbf{y}_i)\}_{i=1}^{m}$, the task is to learn a multi-label classifier

$$h : \mathcal{X} \mapsto \mathcal{Y},$$

where $\mathcal{Y} \subseteq \{-1, 1\}^l$ is the set of feasible label vectors, such that it can predict labels for unseen instances. In practice, instead of learning h directly, it is often to learn a vector-valued function $f : \mathcal{X} \mapsto \mathbb{R}^l$ which determines the label of $\mathbf{x}$ as

$$\hat{\mathbf{y}} = \underset{\mathbf{y}' \in \mathcal{Y}}{\operatorname{argmax}} \ \mathbf{y}'^{\top} f(\mathbf{x}). \tag{1}$$

We can see that how the argmax can be computed depends on $\mathcal{Y}$; for example, if $\mathcal{Y} = \{-1, 1\}^l$, then $\hat{\mathbf{y}}$ can be obtained as $\operatorname{sign}[f(\mathbf{x})]$.

In multi-label learning, a large variety of multi-label performance measures are *example-based performance measures*, which first quantify the performance on each example and then average them over all the examples as the final result. Generally speaking, these performance measures are in the following two groups:

- **Set based performance measures** which evaluate the performance based on the label set prediction of each example, and its representative examples include Hamming loss, F1-score, etc.;
- **Ranking based performance measures** which are based on the ranking of each label for each example, for example, ranking loss and coverage fall in this group.

Without loss of generality, we denote the concerned performance measure as the risk function $\Delta(\mathbf{y}, f(\mathbf{x}))$, which is the smaller the better. For performance measure which is the larger the better, like F1-score, Δ is simply set to one minus it. Obviously, it will be ideal if the classifier f can minimize the expected risk $\mathbb{E}_{(\mathbf{x},\mathbf{y})}[\Delta(\mathbf{y}, f(\mathbf{x}))]$.

2.1   The Problem

In general, multi-label ensemble methods construct a set of multi-label classifiers $\{h^{(t)} : \mathcal{X} \mapsto \mathcal{Y}\}_{t=1}^{k}$, and combine them to produce the vector-valued function $f : \mathcal{X} \mapsto \mathbb{R}^l$ as

$$f(\mathbf{x}; \mathbf{w}) = \sum_{t=1}^{k} w_t h^{(t)}(\mathbf{x}), \qquad (2)$$

where $\mathbf{w} = [w_1, \ldots, w_k]^\top$ is the weighting vector, for example, they are simply set to $1/k$ for voting. It can be found that in (2) the classifier $h^{(t)}$ will be excluded from the ensemble if w_t is zero, and the size of the ensemble is simply $\|\mathbf{w}\|_0$. Then, the task of multi-label selective ensemble becomes to find a weighting vector $\mathbf{w}$, such that the expected performance in terms of the concerned performance measure is optimized, while the ensemble size $\|\mathbf{w}\|_0$ is small.

Since the expectation is infeasible, empirical risk is often used, and the problem of multi-label selective ensemble can be written as

$$\min_{\mathbf{w} \in \mathcal{W}} \frac{1}{m} \sum_{i=1}^{m} \Delta(\mathbf{y}_i, f(\mathbf{x}_i; \mathbf{w})) \quad \text{s.t.} \ \|\mathbf{w}\|_0 \leq b, \qquad (3)$$

where $\mathcal{W}$ is the feasible space of $\mathbf{w}$, and $0 < b \leq k$ is the budget of ensemble size. However, this problem is challenging to solve, mainly because the risk function Δ is among various multi-label performance measures which are generally non-decomposable over labels, non-convex and non-smooth.

2.2   A Convex Formulation

Inspired by the works on structured prediction [23], instead of directly optimizing the empirical risk, in MUSE, we consider to optimize one of its convex upper bounds. Before giving the upper bound, we first make a definition.

Definition 1. *A rank vector* $\mathbf{r}$ *is a permutation of the integer vector* $[1, 2, \ldots, l]$, *and the rank vector* $\mathbf{r}$ *is said to be consistent with the label vector* $\mathbf{y}$, *if and only if there does not exist an index pair* $\langle i, j \rangle$ *satisfying* $y_i = 1$, $y_j = -1$, *and* $r_i < r_j$.

In practice, given a multi-label prediction $\mathbf{p} = f(\mathbf{x})$, the corresponding rank vector can be obtained by sorting $\mathbf{p}$ in ascending order, i.e., if p_t is the smallest in $\mathbf{p}$ then r_t is 1, and the largest corresponds to l. A rank vector $\mathbf{r}$ is consistent with the label vector $\mathbf{y}$, if all the relevant labels indicated by $\mathbf{y}$ have larger rank value than non-relevant ones.

Proposition 1. *Given a multi-label classifier* $f : \mathcal{X} \mapsto \mathbb{R}^l$ *and a multi-label performance measure* Δ,

(a) if Δ *is a set based performance measure, define the loss function*

$$\ell(\mathbf{y}, f(\mathbf{x})) = \max_{\mathbf{y}' \in \mathcal{Y}} \left[(\mathbf{y}' - \mathbf{y})^\top f(\mathbf{x}) + \Delta(\mathbf{y}, \mathbf{y}') \right], \qquad (4)$$

where $\mathcal{Y}$ *is the set of feasible label vectors,*

(b) if Δ is a ranking based performance measure, define the loss function

$$\ell(\mathbf{y}, f(\mathbf{x})) = \max_{\mathbf{r}' \in \Omega} \left[(\mathbf{r}' - \mathbf{r})^\top f(\mathbf{x}) + \Delta(\mathbf{y}, \mathbf{r}') \right], \tag{5}$$

where $\mathbf{r}$ is a rank vector consistent with $\mathbf{y}$ and Ω is the set of possible rank vectors, then the loss function $\ell(\mathbf{y}, f(\mathbf{x}))$ provides a convex upper bound over $\Delta(\mathbf{y}, f(\mathbf{x}))$.

Proof. It is obvious that the function $\ell(\mathbf{y}, f(\mathbf{x}))$ is convex in f, because it is pointwise maximum of a set of linear functions. For set based performance measure, let $\hat{\mathbf{y}} = \text{sign}[f(\mathbf{x})]$ which is the maximizer of $\mathbf{y}^\top f(\mathbf{x})$ in $\mathcal{Y}$, we can get

$$\ell(\mathbf{y}, f(\mathbf{x})) \geq \hat{\mathbf{y}}^\top f(\mathbf{x}) - \mathbf{y}^\top f(\mathbf{x}) + \Delta(\mathbf{y}, \hat{\mathbf{y}}) \geq \Delta(\mathbf{y}, \hat{\mathbf{y}}).$$

For set based performance measures, it holds $\Delta(\mathbf{y}, \hat{\mathbf{y}}) = \Delta(\mathbf{y}, f(\mathbf{x}))$, thus it is an upper bound.

With respect to ranking based performance measure, let $\hat{\mathbf{r}}$ be the rank vector determined by $f(\mathbf{x})$, it is easy to find that $\hat{\mathbf{r}}$ is the maximizer of $\mathbf{r}^\top f(\mathbf{x})$, and then

$$\ell(\mathbf{y}, f(\mathbf{x})) \geq \hat{\mathbf{r}}^\top f(\mathbf{x}) - \mathbf{r}^\top f(\mathbf{x}) + \Delta(\mathbf{y}, \hat{\mathbf{r}}) \geq \Delta(\mathbf{y}, \hat{\mathbf{r}}).$$

Based on above, we can get the conclusion.   ■

For the optimization problem in (3), by replacing Δ with its upper bound ℓ, and $\|\mathbf{w}\|_0$ with its continuous relaxation $\|\mathbf{w}\|_1$, we obtain the optimization problem of multi-label selective ensemble as

$$\min_{\mathbf{w}} \sum_{i=1}^{m} \ell(\mathbf{y}_i, H_i \mathbf{w}) + \lambda \|\mathbf{w}\|_1, \tag{6}$$

where

$$H_i = [h^{(1)}(\mathbf{x}_i), \cdots, h^{(k)}(\mathbf{x}_i)] \in \mathbb{R}^{l \times k} \tag{7}$$

is the matrix collecting the predictions of $\{h^{(t)}\}_{t=1}^{k}$ on instance $\mathbf{x}_i$, and λ is the regularization parameter trading off the empirical risk and the sparsity of $\mathbf{w}$. Obviously, this is an ℓ_1-regularized convex optimization problem, and we solve it via stochastic optimization subsequently.

2.3   Stochastic Optimization

To solve the ℓ_1-regularized convex problem (6), we employ the state-of-the-art stochastic optimization algorithm presented in [21], and the key is how to compute the subgradient of the loss function $\ell(\mathbf{y}_i, H_i \mathbf{w})$.

Proposition 2. Given an example $(\mathbf{x}_i, \mathbf{y}_i)$, a set of multi-label classifiers $\{h^{(t)}\}_{t=1}^{k}$, a weighing vector $\mathbf{w}_0 \in \mathbb{R}^k$ and a multi-label performance measure Δ, denote

$$\mathbf{p}_i = H_i \mathbf{w}_0$$

be the ensemble's prediction on example $(\mathbf{x}_i, \mathbf{y}_i)$ with H_i defined in (7)

Algorithm 1. Stochastic optimization algorithm for MUSE

Input: training data $S = \{(\mathbf{x}_i, \mathbf{y}_i)\}_{i=1}^m$
       component classifiers $\{h^{(t)}\}_{t=1}^k$
       performance measure $\Delta(\cdot, \cdot)$
       regularization parameter λ, step size η
Procedure:
1: let $\mathbf{w} = 0$, $\varrho = 0$ and $p = 2\ln k$
2: **repeat**
3:     select $(\mathbf{x}_i, \mathbf{y}_i)$ uniformly at random from S
4:     let $\mathrm{H}_i = [h^{(1)}(\mathbf{x}_i), \cdots, h^{(k)}(\mathbf{x}_i)]$ and $\mathbf{p}_i = \mathrm{H}_i \mathbf{w}$
5:     solve the argmax problem, *i.e.*,
        $\tilde{\mathbf{y}} \leftarrow$ solve (8) for set based performance measure Δ, or
        $\tilde{\mathbf{r}} \leftarrow$ solve (9) for ranking based performance measure Δ
6:     compute the sub-gradient, *i.e.*,
        $\mathbf{g} = (\tilde{\mathbf{y}} - \mathbf{y}_i)^\top \mathrm{H}_i$ for set based performance measure Δ, or
        $\mathbf{g} = (\tilde{\mathbf{r}} - \mathbf{r}_i)^\top \mathrm{H}_i$ for ranking based performance measure Δ
7:     let $\tilde{\varrho} = \varrho - \eta\mathbf{g}$
8:     let $\forall t$, $\varrho_t = \mathrm{sign}(\tilde{\varrho}_t)\max(0, |\tilde{\varrho}_t| - \eta\lambda)$
9:     let $\forall t$, $w_t = \mathrm{sign}(\varrho_t)|\varrho_t|^{p-1}/\|\varrho\|_p^{p-2}$
10: **until** convergence
Output: weighting vector $\mathbf{w}$

(a) if Δ is a set based performance measure, let $\mathbf{g} = (\tilde{\mathbf{y}} - \mathbf{y}_i)^\top \mathrm{H}_i$, where

$$\tilde{\mathbf{y}} = \underset{\mathbf{y}' \in \mathcal{Y}}{\mathrm{argmax}} \left[\mathbf{y}'^\top \mathbf{p}_i + \Delta(\mathbf{y}_i, \mathbf{y}') \right], \tag{8}$$

(b) if Δ is a ranking based performance measure, let $\mathbf{g} = (\tilde{\mathbf{r}} - \mathbf{r}_i)^\top \mathrm{H}_i$, where $\mathbf{r}$ is a rank vector consistent with $\mathbf{y}_i$ and

$$\tilde{\mathbf{r}} = \underset{\mathbf{r}' \in \Omega}{\mathrm{argmax}} \left[\mathbf{r}'^\top \mathbf{p}_i + \Delta(\mathbf{y}, \mathbf{r}') \right], \tag{9}$$

then the vector $\mathbf{g}$ is a subgradient of $\ell(\mathbf{y}_i, \mathrm{H}_i \mathbf{w})$ at $\mathbf{w}_0$.

Proof. Since $\ell(\mathbf{y}_i, \mathrm{H}_i \mathbf{w})$ is a pointwise maximum of linear functions in $\mathbf{w}$, it is straightforward to obtain its subgradient if the maximizer of (4) or (5) at $\mathbf{w}_0$ can be obtained. Obviously, the argmax (8) and (9) solve the maximizers for set and ranking based performance measures respectively, which completes the proof. ∎

This proposition provides a method to compute the subgradient of $\ell(\mathbf{y}_i, \mathrm{H}_i \mathbf{w})$. Based on this, we can present the stochastic optimization method for solving the optimization problem (6), which is summarized in Algorithm 1. At each iteration, this algorithm first samples an example $(\mathbf{x}_i, \mathbf{y}_i)$ uniformly at random from data S, and then compute the subgradient of $\ell(\mathbf{y}_i, \mathrm{H}_i \mathbf{w})$ (lines 4-6). Since the example $(\mathbf{x}_i, \mathbf{y}_i)$ is chosen at random, the vector $\mathbf{g}$ is an unbiased estimate of the gradient of the empirical risk $\sum_{i=1}^m \ell(\mathbf{y}_i, \mathrm{H}_i \mathbf{w})$. Next, the dual vector ϱ is updated with step size η (line 7) so that the empirical risk is decreased; and also

---

Algorithm 2. Solve the argmax in (9) for coverage

Input: true label vector **y**, current prediction **p**
Procedure:
1: let $max\_val = -\inf$
2: **for** $t \in \{t \mid y_t = 1\}$ **do**
3:     let $\mathbf{v} = \mathbf{p} - \mathbf{e}_t$
4:     $\mathbf{r} \leftarrow$ obtain rank vector by sorting $\mathbf{v}$ ascendingly
5:     **if** $\mathbf{r}^\top \mathbf{v} > max\_val$ **then**
6:         let $\tilde{\mathbf{r}} = \mathbf{r}$ and $max\_val = \mathbf{r}^\top \mathbf{v}$
7:     **end if**
8: **end for**
Output: rank vector $\tilde{\mathbf{r}}$

---

it is truncated to decrease the regularizer $\lambda \|\mathbf{w}\|_1$ (line 8). Finally, the updates of ϱ is translated to the variable $\mathbf{w}$ via a link function in line 9. This procedure iterates until convergence.

Solving the Argmax. In order to make Algorithm 1 practical, the argmax in (8) and (9) need to be solved for the concerned performance measure. Fortunately, there have been proposed efficient procedures for many commonly used performance measures.

For examples, Joachims [12] solved the argmax problem in $O(l^2)$ time for a large class of set based measures including Hamming loss and F1-score, in $O(l \log l)$ time for ranking loss; Yue *et al.* [29] solved it for average precision in $O(l \log l)$ time; Le *et al.* [13] solved (9) by a linear assignment problem for a group of ranking based performance measures including precision@k. Of course, if our concerned performance measure is among them, these procedures can be directly employed by Algorithm 1. Here, we omit the detailed procedures, which can be found in [12,13,29].

To our best knowledge, there is still no proposal to solve the argmax (9) for *coverage*, which is ranking based performance measure evaluating how far one needs to go along the list of labels to cover all the true labels. Formally, given a true label vector **y** and a rank vector **r**, it is defined as

$$\Delta_c(\mathbf{y}, \mathbf{r}) = \max_{\{t \mid y_t = 1\}} (l - r_t) = \max_{t \in \{t \mid y_t = 1\}} (l - \mathbf{r}^\top \mathbf{e}_t), \tag{10}$$

where $\mathbf{e}_t$ is a vector with t-th element as 1 and others 0. Substituting (10) into the argmax problem (9), we can get an equivalent problem as

$$\tilde{\mathbf{r}} = \operatorname*{argmax}_{\mathbf{r}' \in \Omega} \max_{t \in \{t \mid y_t = 1\}} \mathbf{r}'^\top (\mathbf{p}_i - \mathbf{e}_t). \tag{11}$$

It is not difficult to see that for one single t, the problem

$$\operatorname*{argmax}_{\mathbf{r}' \in \Omega} \mathbf{r}'^\top (\mathbf{p}_i - \mathbf{e}_t) \tag{12}$$

can be efficiently solved by sorting the vector $(\mathbf{p}_i - \mathbf{e}_t)$ ascendingly, and obtaining the corresponding rank vector. As a consequence, by enumerating all t's and solving the corresponding problems as (12), the solution to (11) can be obtained. The pseudocode of this procedure is given in Algorithm 2. We can find that in each iteration, the complexity is dominated by the sorting in line 4, thus the total complexity of Algorithm 2 is $O(sl \log l)$, where s is the number of true labels of current example.

Convergence and Computational Complexity. Based on Theorem 3 in [21], we can find that the number of iterations of Algorithm 1 to achieve ϵ-accuracy is bounded by $O(\log k/\epsilon^2)$ with k as the number of component classifiers. It can be found that this number is independent of the data size. Moreover, in each iteration, all the operations are performed on one single example, and the complexity is dominated by the argmax which as shown above can be solved in polynomial time for various performance measures, also independent of data size. This constitutes one of the appealing properties of MUSE, i.e., at each iteration of Algorithm 1, we neither need to compute the predictions on all examples nor need to solve the argmax on all examples.

3   Experiments

In this section, we perform a set of experiments to evaluate the effectiveness of our proposed MUSE approach.

3.1   Configuration

The experiments are performed on image and music annotation tasks. Specifically, two image annotation tasks are used, including *corel5k* which has 5000 images and 374 possible labels, and *scene* which has 2407 images and 6 possible labels; two music annotation tasks are used, including *cal500* which has 502 songs and 174 possible labels, and *emotion* which has 593 songs and 6 possible labels. Five representative multi-label performance measures are considered for each task, including Hamming loss, precision@k, F1-score, coverage, ranking loss. The formal definition of these performance measures can be found in [20], and the k of precision@k is set to the average number of relevant labels. In total, there are 20 tasks.

In experiments, MUSE is implemented based on ECC [19], that is, by using ECC, we first obtain 100 classifier chains, then use MUSE to obtain the selective ensemble out of them. We compare MUSE with BSVM [1] which trains one SVM for each label, and state-of-the-art methods including the lazy method ML-kNN [31], label ranking method CLR [6] and the full ECC combining all classifiers. It is also compared with the *random strategy* which selects classifiers randomly. Specifically, LibLinear [5] is used to implement the base classifier in BSVM and ECC; the default implementation of ML-kNN and CLR in Mulan [24] are used; the random strategy generates ensembles of the same size of MUSE.

For each task, these comparative methods are evaluated by using 30 times random holdout test, i.e., 2/3 for training and 1/3 for testing in each time; finally, averaged performance and standard derivation are reported; the sizes of the generated selective ensembles are reported. For MUSE, the regularization parameter λ is chosen by 5-fold cross validation on training set.

3.2   Results

Optimizing the Upper Bound. Instead of optimizing the concerned performance measures directly, the proposed MUSE approach tries to optimize its convex upper bound. Thus, a natural question is whether this is effective, or in other words, whether optimizing the upper bound will improve the performance. To answer this question, we record the training and test performance on *cal500*, and the results are shown in Fig. 2. It can be found that both the training and test performance improve when the optimization procedure goes on, which give a positive answer to above question, that is, optimizing the upper bound is effective in improving the performance. Moreover, we can see from Fig. 2 that the performance converges after some iterations. For example, they converge after about 600 iterations for coverage. Noting there are 502 examples in total, and in each iteration, MUSE operates on only one example, this means that we need to scan the data set for only once.

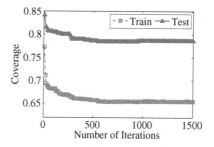

Fig. 2. Both the training and test performance improve during the optimization procedure, where coverage is evaluated on the cal500 data set.

Performance Comparison. The performance of all the comparative methods are shown in Table 1. For better comparison, we perform paired t-tests at 95 % significance level to compare MUSE with other methods, and the results are also shown in Table 1.

It can be seen that the performance of MUSE is quite promising. Compared with the full ensemble ECC_{100}, it achieves 3 wins and 16 ties and loses only 1 time out of all 20 tasks, while the ensemble size is reduced. For example, on *cal500* the F1-score is improved from 0.323 to 0.384, but the ensemble size reduced from 100 to less than 20. Also, when compared with the random strategy, MUSE achieves 9 wins but 0 loss, which shows its effectiveness. Comparing MUSE with other methods, we can see that it achieves significantly better performance (15 wins

Table 1. Experimental results (mean±std.), where ●(○) indicates that MUSE is significantly better (worse) than the corresponding method based on paired t-tests at 95 % significance level, the sizes of selective ensembles generated by MUSE are reported (after '/'), and the win/tie/loss counts based on paired t-tests are summarized in the last row. Note that the ensemble size of ECC_{100} is 100, and the random strategy generates ensembles of the same size of MUSE.

Data	BSVM	ML-kNN	CLR	ECC_{100}	Random	MUSE
Hamming loss (the smaller, the better)						
corel5k	.014±.001●	.009±.001	.498±.003●	.010±.001	.010±.002	.010±.001/43.9±12.2
scene	.120±.003●	.089±.004○	.174±.004●	.102±.003	.180±.007●	.101±.003/66.2±18.4
cal500	.212±.012●	.139±.002	.378±.002●	.141±.002	.145±.005●	.137±.002/62.9± 7.4
emotion	.305±.011	.201±.012○	.190±.012○	.302±.012	.312±.012●	.297±.012/36.3±15.9
Precision@k (the larger, the better)						
corel5k	.177±.005●	.213±.005●	.232±.005	.234±.006	.227±.009	.231±.006/63.3±12.6
scene	.451±.006●	.472±.006	.453±.007●	.465±.005	.460±.013●	.469±.001/83.0±7.8
cal500	.315±.025●	.447±.006	.453±.006	.452±.007	.438±.010	.446±.007/74.3± 7.2
emotion	.605±.017	.618±.020	.582±.016●	.609±.022	.603±.022	.607±.021/53.2±20.1
F1-score (the larger, the better)						
corel5k	.135±.005●	.016±.003●	.034±.001●	.134±.006●	.127±.008●	.149±.006/15.3± 5.6
scene	.595±.014●	.675±.018	.629±.009●	.668±.012	.567±.014●	.672±.002/22.0±10.8
cal500	.314±.029●	.322±.011●	.405±.003○	.323±.010●	.333±.023●	.384±.013/19.6± 4.7
emotion	.614±.014	.602±.029●	.622±.016	.618±.015	.612±.016	.619±.017/63.7±21.2
Coverage (the smaller, the better)						
corel5k	.560±.008●	.309±.004○	.285±.005○	.362±.007	.367±.011	.363±.011/74.1±13.7
scene	.102±.004	.090±.004	.102±.006	.096±.004	.097±.008	.097±.005/76.3±13.2
cal500	.913±.015●	.750±.010○	.756±.008○	.768±.013○	.796±.002●	.790±.002/41.8± 9.7
emotion	.322±.015	.319±.013	.323±.014	.328±.015	.334±.019	.326±.017/73.6±12.5
Ranking loss (the smaller, the better)						
corel5k	.270±.005●	.135±.002●	.119±.002●	.042±.001●	.039±.005●	.032±.003/71.3±11.7
scene	.106±.005●	.082±.005	.079±.005	.087±.004	.087±.007	.086±.004/52.4± 9.7
cal500	.309±.019●	.184±.003	.182±.003	.184±.005	.179±.005	.188±.006/54.9±12.4
emotion	.194±.012●	.169±.015○	.146±.013○	.187±.013	.187±.016	.188±.011/74.1±11.7
Win/Tie/Loss counts (MUSE vs alternatives)						
counts	15/5/0	5/11/4	8/9/3	3/16/1	9/11/0	–

and 0 loss) against BSVM, also comparable performance against the state-of-the-art methods ML-kNN and CLR.

4 Related Work

In ensemble learning, selective ensemble (*a.k.a.* ensemble pruning or ensemble selection) is an active research topic [34, chapter6]. In traditional supervised learning, a number of methods have been developed based on different techniques, such as genetic algorithm [35], semi-definite programming [33], clustering [9], ℓ_1-norm regularized sparse optimization [14]. In [15], in order to reduce the size of ECC, Li and Zhou proposed SECC (i.e., selective ensemble of classifier chains), which to our best knowledge is the first work on selective ensemble in the multi-label setting.

In this paper, we address the problem of multi-label selective ensemble by proposing that it is needed to take the concerned performance measure fully into account, and propose the MUSE approach to build the selective ensemble via sparse convex optimization. This is encouraged and inspired by recent works on optimizing complicated performance measures [12,15,17]. Moreover, compared with previous work [15], the MUSE approach is more general, for instance, it can optimize a large variety of performance measures while SECC considers only F1-score. In other words, SECC is a special case of MUSE when it considers only F1-score.

5   Conclusion

In this paper, we study the problem of multi-label selective ensemble, which tries to select a subset of component classifiers whilst keeping or improving the performance. The main motivation is that we need to take the concerned performance measure into account during the selection process, and the MUSE approach is proposed to handle this problem. Specifically, by taking an upper bound over empirical risk, MUSE tries to optimize the concerned performance measure via an ℓ_1-norm regularized convex optimization problem. And this problem can be efficiently solved by stochastic subgradient descend for a large variety of performance measures. Experiments on image and music annotation tasks show the effectiveness of the proposed MUSE approach.

In current work, we consider the component classifier in multi-label ensemble as a general multi-label classifier. Often, the component classifier itself is a group of single-label classifiers, like classifier chain in ECC. Therefore, it will be interesting to consider the multi-label selective ensemble problem at the level of such single-label classifiers.

Acknowledgements. We want to thank anonymous reviewers for helpful comments. This research was supported by the National Science Foundation of China (61273301).

References

1. Boutell, M., Luo, J., Shen, X., Brown, C.: Learning multi-label scene classification. Pattern Recogn. **37**(9), 1757–1771 (2004)
2. Bucak, S.S., Jin, R., Jain, A.: Multi-label learning with incomplete class assignments. In Proceedings of the IEEE Computer Society Conference on Computer Vision and Pattern Recognition, Colorado Springs, CO, pp. 2801–2808 (2011)
3. Dembczynski, K., Cheng, W., Hüllermeier, E.: Bayes optimal multilabel classification via probabilistic classifier chains. In: Proceedings of the 27th International Conference on Machine Learning, Haifa, Israel, pp. 279–286 (2010)
4. Dembczynski, K., Waegeman, W., Cheng, W., Hüllermeier, E.: Regret analysis for performance metrics in multi-label classification. In: Proceedings of the 21st European Conference on Machine Learning, Barcelona, Spain, pp. 280–295 (2010)
5. Fan, R.-E., Chang, K.-W., Hsieh, C.-J., Wang, X.-R., Lin, C.-J.: Liblinear: a library for large linear classification. J. Mach. Learn. Res. **9**, 1871–1874 (2008)

6. Fürnkranz, J., Hüllermeier, E., Mencía, E.L., Brinker, K.: Multilabel classification via calibrated label ranking. Mach. Learn. **73**(2), 133–153 (2008)
7. Gao, W., Zhou, Z.-H.: On the consistency of multi-label learning. Artif. Intell. **199–200**, 22–44 (2013)
8. Ghamrawi, N., McCallum, A.: Collective multi-label classification. In: Proceedings of the 14th ACM International Conference on Information and Knowledge Management, Bremen, Germany, pp. 195–200 (2005)
9. Giacinto, G., Roli, F., Fumera, G.: Design of effective multiple classifier systems by clustering of classifiers. In: Proceedings of the 15th International Conference on Pattern Recognition, Barcelona, Spain, pp. 160–163 (2000)
10. Hariharan, B., Zelnik-Manor, L., Vishwanathan, S., Varma, M.: Large scale max-margin multi-label classification with priors. In: Proceedings of the 27th International Conference on Machine Learning, Haifa, Israel, pp. 423–430 (2010)
11. Hsu, D., Kakade, S., Langford, J., Zhang, T.: Multilabel prediction via compressed sensing. In: Advances in Neural Information Processing Systems 22, pp. 772–780. MIT Press, Cambridge (2009)
12. Joachims, T.: A support vector method for multivariate performance measures. In: Proceedings of the 22nd International Conference on Machine Learning, Bonn, Germany, pp. 377–384 (2005)
13. Le, Q., Smola, A.: Direct optimization of ranking measures (2007). CoRR. abs/0704.3359
14. Li, N., Zhou, Z.-H.: Selective ensemble under regularization framework. In: Benediktsson, J.A., Kittler, J., Roli, F. (eds.) MCS 2009. LNCS, vol. 5519, pp. 293–303. Springer, Heidelberg (2009)
15. Li, N., Zhou, Z.-H.: Selective ensemble of classifier chains. In: Zhou, Z.-H., Roli, F., Kittler, J. (eds.) MCS 2013. LNCS, vol. 7872, pp. 146–156. Springer, Heidelberg (2013)
16. McCallum, A.: Multi-label text classification with a mixture model trained by EM. In: Working Notes of AAAI 1999 Workshop on Text Learning (1999)
17. Nan, Y., Chai, K.M., Lee, W., Chieu, H.: Optimizing F-measure: a tale of two approaches. In: Proceedings of the 29th International Conference on Machine Learning, Edinburgh, UK, pp. 289–296 (2012)
18. Read, J., Pfahringer, B., Holmes, G.: Multi-label classification using ensembles of pruned sets. In: Proceedings of the 8th IEEE International Conference on Data Mining, Pisa, Italy, pp. 995–1000 (2008)
19. Read, J., Pfahringer, B., Holmes, G., Frank, E.: Classifier chains for multi-label classification. Mach. Learn. **85**(3), 333–359 (2011)
20. Schapire, R., Singer, Y.: BoosTexter: a boosting-based system for text categorization. Mach. Learn. **39**(2–3), 135–168 (2000)
21. Shalev-Shwartz, S., Tewari, A.: Stochastic methods for ℓ_1-regularized loss minimization. J. Mach. Learn. Res. **12**, 1865–1892 (2011)
22. Shi, C., Kong, X., Yu, P., Wang, B.: Multi-label ensemble learning. In Proceedings of the 22nd European Conference on Machine learning, Athens, Greece, pp. 223–239 (2011)
23. Tsochantaridis, I., Joachims, T., Hofmann, T., Altun, Y.: Large margin methods for structured and interdependent output variables. J. Mach. Learn. Res. **6**, 1453–1484 (2005)
24. Tsoumakas, G., Spyromitros-Xioufis, E., Vilcek, J., Vlahavas, I.: MULAN: a Java library for multi-label learning. J. Mach. Learn. Res. **12**, 2411–2414 (2011)

25. Tsoumakas, G., Vlahavas, I.: Random k-labelsets: an ensemble method for multi-label classification. In: Proceedings of the 18th European Conference on Machine Learning, Warsaw, Poland, pp. 406–417 (2007)
26. Turnbull, D., Barrington, L., Torres, D., Lanckriet, G.: Semantic annotation and retrieval of music and sound effects. IEEE Trans. Audio Speech Lang. Process. **16**(2), 467–476 (2008)
27. Ueda, N., Saito, K.: Parametric mixture models for multi-labeled text. In: Advances in Neural Information Processing Systems 15, pp. 721–728. MIT Press, Cambridge (2003)
28. Xu, M., Li, Y.-F., Zhou, Z.-H.: Multi-label learning with pro loss. In: Proceedings of the 27th AAAI Conference on Artificial Intelligence, Bellevue, WA, pp. 998–1004 (2013)
29. Yue, Y., Finley, T., Radlinski, F., Joachims, T.: A support vector method for optimizing average precision. In: Proceedings of the 30th Annual International ACM SIGIR Conference on Research and Development in Information Retrieval, Amsterdam, Netherlands, pp. 271–278 (2007)
30. Zhang, M.-L., Zhang, K.: Multi-label learning by exploiting label dependency. In: Proceedings of the 16th ACM SIGKDD Conference on Knowledge Discovery and Data Mining, Washington, DC, pp. 999–1007 (2010)
31. Zhang, M.-L., Zhou, Z.-H.: ML-KNN: a lazy learning approach to multi-label learning. Pattern Recogn. **40**(7), 2038–2048 (2007)
32. Zhang, M.-L., Zhou, Z.-H.: A review on multi-label learning algorithms. IEEE Trans. Knowl. Data Eng. **26**(8), 1819–1837 (2014)
33. Zhang, Y., Burer, S., Street, W.: Ensemble pruning via semi-definite programming. J. Mach. Learn. Res. **7**, 1315–1338 (2006)
34. Zhou, Z.-H.: Ensemble Methods: Foundations and Algorithms. Chapman and Hall/CRC, Boca Raton (2012)
35. Zhou, Z.-H., Wu, J., Tang, W.: Ensembling neural networks: many could be better than all. Artif. Intell. **137**(1–2), 239–263 (2002)

Supervised Selective Combination of Diverse Object-Representation Modalities for Regression Estimation

Olga Krasotkina[1(✉)], Oleg Seredin[1], and Vadim Mottl[2]

[1] Tula State University, Lenin Ave. 92, 300012 Tula, Russia
{kol80177, oseredin}@yandex.ru
[2] Computing Center of the Russian Academy of Sciences,
Vavilov St., 40, 117968 Moscow, Russia
vmottl@yandex.ru

Abstract. We consider the problem of multi-modal regression estimation under the assumption that a kernel-based approach is applicable within each particular modality. The Cartesian product of the linear spaces into which the respective kernels embed the output scales of single sensors is employed as an appropriate joint scale corresponding to the idea of combining modalities at the sensor level. This contrasts with the commonly adopted method of combining classifiers inferred from each specific modality. However, a significant risk in combining linear spaces is that of overfitting. To address this, we set out a stochastic method for encompassing modal-selectivity that is intrinsic to (that is to say, theoretically contiguous with) the selected kernel-based approach.

Keywords: Kernel-based regression · Combining modalities · Kernel fusion · Classifier fusion

1 Introduction

Problems of estimating dependencies from empirical data belong to the most glowing challenges of modern informatics. Let $\omega \in \Omega$ be a set of real-world objects naturally associated with a hidden characteristic $y \in \mathbb{Y}$. The function $y(\omega) : \Omega \to \mathbb{Y}$ is known to the observer only within the bounds of a finite training set

$$\Omega^* \Rightarrow \big\{ \big(\omega_j, y(\omega_j)\big), j = 1, \ldots, N \big\}. \tag{1}$$

It is required to continue the function onto the entire set $\hat{y}(\omega) : \Omega \to \mathbb{Y}$, so that it would be possible to estimate the values of the goal characteristic for other objects $\omega \in \Omega \backslash \Omega^*$ [1]. This scenario of precedent-based dependence estimation is said to be the problem of pattern recognition if the hidden function takes values from a finite set $\hat{y}(\omega) : \Omega \to \{y^{(1)}, \ldots, y^{(m)}\}$, and is referred to as that of regression estimation in the case of the real-valued function $\hat{y}(\omega) : \Omega \to \mathbb{R}$. It is just the latter kind of dependence recovery problems, which is addressed in this paper.

A computer is incapable of immediate perceiving any physical entity, therefore, a formal variable must ever act as the mediator of the real world to it. Practically all

© Springer International Publishing Switzerland 2015
F. Schwenker et al. (Eds.): MCS 2015, LNCS 9132, pp. 89–99, 2015.
DOI: 10.1007/978-3-319-20248-8_8

principles of object representation in designing dependence-estimation techniques boil down into the categories of feature-based and similarity/dissimilarity-based ones.

The feature-based principle associates objects with a variable $x(\omega) : \Omega \to \mathbb{X}$ called their computer-perceptible feature. The unknown regression dependence $\hat{y}(x(\omega)) : \mathbb{X} \to \mathbb{R}$ is to be estimated from a more specific training set than (1):

$$\Omega^* \Rightarrow \left\{ (x(\omega_j), y(\omega_j)), j = 1, \ldots, N \right\}. \tag{2}$$

The simplest assumption which has given rise to the most popular feature-based methods is that the objects are represented by real vectors $x(\omega) = (x_1(\omega), \ldots, x_n(\omega)) \in \mathbb{R}^n$. Then, the regression model $\hat{y}(x(\omega)) : \mathbb{R}^n \to \mathbb{R}$ is particularly simple in both formulation and estimation of parameters $\hat{c} \in \mathbb{R}^n$ and $\hat{b} \in \mathbb{R}$ from the training set (2):

$$\hat{y}(x(\omega)) = \hat{c}^T x(\omega) + \hat{b} = \sum\nolimits_{i=1}^{n} \hat{c}_i x_i(\omega) + \hat{b}, \tag{3}$$

It is to be noticed that single features $x_i(\omega)$ in (3) are, actually, simplest object-representation modalities to be fused by way of their linear combination.

In this paper, we address the alternative and more general similarity/dissimilarity-based principle of object representation, which implies that the only way to perceive real-world objects is pair-wise comparison of them (ω', ω'') by a real two-argument function $K(\omega', \omega'') : \Omega \times \Omega \to \mathbb{R}$. In the majority of practical situations, similarities or distinctions between pairs of objects are to be measured from the viewpoints of several different properties, i.e., object-comparison modalities, each expressed by a modality-specific function $K_i(\omega', \omega'')$, $i = 1, \ldots, n$. In a training set for similarity/dissimilarity-based estimation of the regression dependence, objects must be represented, instead of individual features (2), by n matrices of pair-wise object comparison:

$$\Omega^* \Rightarrow \left\{ K_i(\omega_j, \omega_l), i = 1, \ldots, n, \ y(\omega_j), \ j, l = 1, \ldots, N \right\}. \tag{4}$$

More specifically, we consider here the kernel-based principle of object representation, which assumes that the comparison functions $K_i(\omega', \omega'') : \Omega \times \Omega \to \mathbb{R}$ are kernels in the set of objects [2]. This means that each of them is symmetric $K_i(\omega', \omega'') = K_i(\omega'', \omega')$, and the respective matrix in the training set (4) is always positive-semidefinite. One of the main advantages of the kernel-based approach is its ability to facilitate easily fusing several seemingly incomparable physical properties of objects in an entire model of the dependence of interest $y(\omega)$ [3].

We show that the result of kernel-based regression estimation has, under some natural assumptions, the closed form

$$\hat{y}(\omega|\Omega^*) = \sum\nolimits_{j=1}^{N} \sum\nolimits_{i=1}^{n} \hat{a}_{ij} K_i(\omega_j, \omega) + \hat{b}, \tag{5}$$

where real-valued parameters $(\hat{a}_{ij}, \ i = 1, \ldots, n, \ j = 1, \ldots, N)$ and $\hat{b}$ are to be inferred from the training set (4).

The greater the variety of different modalities, the broader the diversity of object properties which underlie the regression model (5). But if the number of kernels n is too large for the size N of the available training set, the overcomplicated model loses its generalization performance. The problem of selection among real-valued regressors (3) has been broadly considered in the literature [4], but the existing methodology does not cover the kernel-based regression model (5). This paper is endeavoured to partially fill in this gap.

For selective combination of several object-representation kernels when estimating the regression model from the observed training set (4), we adopt the principle of Relevance Kernel Machine (RKM) originally developed for the purpose of pattern recognition [5]. The resulting training technique will be regressor selective in the sense that it is able to find redundant features and assign to them small regression coefficients $(\hat{a}_{ij}, j = 1, \ldots, N)$.

The desired selectivity is achieved through a meta-parameter that controls the model complexity, i.e., the degree of elimination of redundant kernels. The appropriate level of selectivity is determined via cross-validation procedure.

2 A Kernel-Based Parametric Family of Regression Dependencies Over Objects of Arbitrary Kind

We shall assume throughout this paper that n kernel functions $(K_i(\omega', \omega''):$ $\Omega \times \Omega \rightarrow \mathbb{R}, i = 1, \ldots, n)$ are defined in the set of real-world objects of interest $\omega \in \Omega$, which express alternative ways of quantitative comparison between all the pairs of objects. A kernel is a symmetric two argument function that forms a positive semi definite matrix $(K_i(\omega_j, \omega_l), j, l = 1, \ldots, m)$ for any finite collection of objects $\{\omega_j, j = 1, \ldots, m\}$ [6]. Each of the kernels $K_i(\omega', \omega'')$ embeds the same set of objects Ω into a specific into hypothetical linear space $\Omega \subset \tilde{\Omega}_i$, in which the null element and linear operations are defined in a particular way [7]:

$$\phi_i \in \tilde{\Omega}_i, \ \omega' + \omega'' : \tilde{\Omega}_i \times \tilde{\Omega}_i \rightarrow \tilde{\Omega}_i, \ a\omega : \mathbb{R} \times \tilde{\Omega}_i \rightarrow \tilde{\Omega}_i \qquad (6)$$

The role of inner product is played by the symmetric kernel function itself which is inevitably bilinear $K_i(a'\omega' + a''\omega'', \omega) = a'K_i(\omega', \omega) + a''K_i(\omega'', \omega)$.

The major convenience factor of the kernel-based approach to data analysis is its ability to provide the constructor of a data-analysis system with the possibility of working with objects of arbitrary nature in unified terms of linear real-valued functions $f_i(\omega) : \Omega \rightarrow \mathbb{R}$. More strictly, the carrier of kernel-specific linear functions is not the set of objects itself Ω, but rather its respective linear closure $\Omega \subset \tilde{\Omega}_i \rightarrow \mathbb{R}$. To determine a scalar linear function $f_i(\omega) : \tilde{\Omega}_i \rightarrow \mathbb{R}$, it is enough to specify a direction element (vector, in linear-space terms) $c_i \in \tilde{\Omega}_i$, then the function will be expressed as inner product $f_i(\omega \,|\, c_i) = K_i(c_i, \omega)$.

Let us consider now the Cartesian product $\tilde{\Omega}_1 \times \ldots \times \tilde{\Omega}_n \supset \Omega \times \ldots \times \Omega = \Omega^n$ of the linear spaces $\tilde{\Omega}_i \supset \Omega$ defined by the respective kernels, and then assign an appropriate combined kernel (inner product) in it $(\tilde{\Omega}_1 \times \ldots \times \tilde{\Omega}_n) \times (\tilde{\Omega}_1 \times \ldots \times \tilde{\Omega}_n) \to \mathbb{R}$. We shall apply this inner product only to n-fold repetitions of the same real-world object $(\omega, \ldots, \omega) \in \Omega \times \ldots \times \Omega = \Omega^n$ and conventionally use the symbol $K(\omega', \omega'')$ for inner products of such constructions. In particular, the sum of the initial kernels $K(\omega', \omega'') = \sum_{i=1}^n K_i(\omega', \omega'')$ will be a kernel in $\tilde{\Omega}_1 \times \ldots \times \tilde{\Omega}_n$. This idea is following our previous works [8–11].

From this point of view, any choice of a parameters $c = (c_i \in \tilde{\Omega}_i, i = 1, \ldots, n) \in \tilde{\Omega}_1 \times \ldots \times \tilde{\Omega}_n$ and real number $b \in \mathbb{R}$ yields a linear regression dependence in the set of objects

$$\hat{y}(\omega) = K(c, \omega) + b = \sum_{i=1}^n K_i(c_i, \omega) + b, \tag{7}$$

and produces, thereby, a kernel fusion technique.

However, the kernel-based family of regression dependences (7) contains coefficients $c_i \in \tilde{\Omega}_i$ meant to take values from hypothetical kernel-specific linear spaces deriving from the kernel trick, in contrast to the initial set of real-world objects, which is their common subset $\Omega \subset \tilde{\Omega}_i$. Thus, immediate estimation of these coefficients from the given training set (4) is computationally impossible.

Nevertheless, in the next Sect. 3, we outline quite lenient assumptions on the origin of the observed data set (4), under which its hidden regression model (7) can be estimated in the closed form (5) without any loss of generality.

3   A Linear Normal-Gamma Model of the Hidden Kernel-Based Regression Dependence and Its Bayesian Estimation from the Training Set

3.1   Linear Normal Observation Model

In general, we imply the view of the set of pairs $(\omega, y(\omega)) \in \Omega \times \mathbb{R}$ as a probability space. This means that any observed object and its real-valued goal characteristic are to be treated as a random pair of variables in $\Omega \times \mathbb{R}$. But our approach to regression estimation will fully rest on estimating the unknown conditional density $\varphi^*(y \,|\, \omega)$ in $\mathbb{R}$ and abstract away from the still more unknown marginal distribution of ω in Ω.

Let the observer assume the linear regression model $E(y \,|\, \omega; c_1, \ldots, c_n, b)$ with unknown parameters $c_i \in \tilde{\Omega}_i$ (7) instead of the unknown genuine density $\phi^*(y \,|\, \omega)$ and consider the respective conditional parametric family of distribution densities to be normal with unknown observation noise variance $\xi > 0$:

$$\phi(y \,|\, \omega; c_1, \ldots, c_n, \xi) = \left(1/\xi^{1/2}(2\pi)^{1/2}\right) \, \exp\left(-(1/2\xi)\left(y - \sum_{i=1}^n K_i(c_i, \omega) - b\right)^2\right).$$

If, in addition, the random observed values of the goal variable $y_j = y(\omega_j)$ in the training set (4) are considered as depending each only on its object, so, the joint distribution density will be the product

$$\Phi(y_1, \ldots, y_N \mid \omega_1, \ldots, \omega_N; c_1, \ldots, c_n, \xi) = \prod_{j=1}^{N} \varphi(y_j \mid \omega_j; c_1, \ldots, c_n, \xi)$$

$$= \left(1/\xi^{N/2}(2\pi)^{N/2}\right) \exp\left(-(1/2\xi) \sum_{j=1}^{N} \left(y_j - \sum_{i=1}^{n} K_i(c_i, \omega_j) - b\right)^2\right). \tag{8}$$

3.2 Bayesian Estimation with Fixed a Priori Variances of Kernel-Specific Regression Coefficients

In its turn, the unknown regression coefficients $(c_1, \ldots, c_n)$ are a priori considered as independent hidden random variables distributed in respective kernel-specific linear spaces $c_i \in \tilde{\Omega}_i$ in accordance with zero-mean normal-like circular laws $E(c_i) = \phi_i \in \tilde{\Omega}_i$. Let the circular variances $r_i \xi$ be proportional to the observation noise variance ξ in (8) and individual for each modality due to individual proportionality coefficients $(r_1 > 0, \ldots, r_n > 0)$, so that $\psi_i(c_i \mid r_i, \xi) \propto \left(1/(r_i\xi)^{1/2}\right) \exp(-(1/2r_i\xi) K_i(c_i, c_i)).$[1] As to the random regression constant $b \in \mathbb{R}$, no a priori information is assumed to be available on its distribution. So, the joint a priori density will be expressed in the improper form as the product

$$\Psi(c_1, \ldots, c_n, b \mid r, \ldots, r_n, \xi) \propto \left(\prod_{i=1}^{n} r_i\xi\right)^{-1/2} \exp\left(-(1/2) \sum_{i=1}^{n} (1/r_i\xi) K(c_i, c_i)\right). \tag{9}$$

Let us assume first that the variances of regression coefficients $(r_1\xi, \ldots, r_n\xi)$ in the a priori distribution (9) are fixed. Then, the maximum point of the joint a posteriori density $P(c_1, \ldots, c_n, b \mid \Omega^*, r_1, \ldots, r_n, \xi)$ will be the object of Bayesian training from the given training set (4):

$$(\hat{c}_1, \ldots, \hat{c}_n, \hat{b}) = \arg\max P(c_1 \in \tilde{\Omega}_1, \ldots, c_n \in \tilde{\Omega}_n, b \mid \Omega^*, r_1, \ldots, r_n, \xi)$$
$$= \arg\max[\ln \Phi(y_1, \ldots, y_N \mid \omega_1, \ldots, \omega_N; c_1, \ldots, c_n, \xi) \tag{10}$$
$$+ \ln \Psi(c_1, \ldots, c_n, b \mid r, \ldots, r_n, \xi)].$$

It is immediately seen from (8) and (9) that the estimate will not depend on the assumed observation noise variance ξ. The following Theorem 1 shows that the result of training can be expressed in the explicit form (5) omitting the necessity of solving the optimization problem (10) in terms of hypothetical linear spaces $c_i \in \tilde{\Omega}_i \supset \Omega$.

Theorem 1. The regression model (7) inferred from the training set Ω^* under the a priori assumption of fixed regression coefficient in accordance with the Bayesian

---

[1] It is incorrect to speak about strictly normal densities since the dimensionality of each linear space $\tilde{\Omega}_i$ depends on the respective kernel function. As a result, the normalization coefficient of any density $(\psi_i(a_i), a_i \in \tilde{\Omega}_i)$ cannot be specified before the kernel is completely defined.

condition (10) has the closed form (5) with parameters $(\hat{a}_{ij}, \hat{b})$, which are the minimum point of the quadratic optimization criterion

$$
\begin{aligned}
J(a_{ij}, i &= 1, \ldots, n, b, j = 1, \ldots, N \mid r_i, i = 1, \ldots, n) \\
&= \sum_{i=1}^{n} (1/r_i) \sum_{j=1}^{N} \sum_{l=1}^{N} K_i(\omega_j, \omega_l) a_{ij} a_{il} \\
&+ \sum_{j=1}^{N} \left(y_j - \sum_{i=1}^{n} \sum_{l=1}^{N} K_i(\omega_j, \omega_l) a_{il} - b \right)^2 \rightarrow \min.
\end{aligned}
\tag{11}
$$

In its turn, the minimum point $(\hat{a}_{ij} = r_i \hat{\delta}_j, \hat{b})$ is completely defined by the solution $(\hat{\delta}_1, \ldots, \hat{\delta}_N, \hat{b})$ of the system of $N + 1$ linear equations

$$
\begin{cases}
\left(\sum_{i=1}^{n} r_i K_i(\omega_j, \omega_j) + 1 \right) \delta_j + \sum_{l=1, l \neq j}^{N} \left(\sum_{i=1}^{n} r_i K_i(\omega_j, \omega_l) \right) \delta_l + b = y_j, & j = 1, \ldots, N, \\
\sum_{j=1}^{N} \delta_j = 0,
\end{cases}
\tag{12}
$$

and results in the Bayesian estimate of the regression model (5)

$$
\hat{y}(\omega \mid \Omega^*, r_1, \ldots, r_n) = \sum_{j=1}^{N} \hat{\delta}_j \sum_{i=1}^{n} r_i K_i(\omega_j, \omega) + \hat{b}.
\tag{13}
$$

It is well seen from (13) that the assumed positive coefficients $(r_1, \ldots, r_n)$ occur in the estimated regression model as weights at the object-representation kernels. If all the coefficients equal unity $r_1 = \ldots = r_n = 1$, all the kernels equally participate in the model, but when it is required to suppress some of them, it is enough to take the respective coefficients r_i close to zero, then so will be also the a priori variances $r_i \xi$ of hypothetical regression coefficients a_i in (9).

Further maximization of the Bayesian criterion (10) also with respect to $(r_1, \ldots, r_n)$ is senseless, because it will ever prefer greater values of all coefficients. The coefficients $(r_1, \ldots, r_n)$ play the role of a vector structural parameter of the class of regression models, and their appropriate values are to be chosen from the requirement of maximum generalization performance of training from a single sample set Ω^*.

In the next Sect. 3.3, we consider the unknown reciprocal variance coefficients $1/r_i$, in their turn, as independent random variables a priori distributed in accordance with identical gamma densities $\gamma((1/r_i) \mid \mu) \propto (1/r_i)^{((1+\mu)^2/2\mu) - 1} \exp(-(1/2\mu)(1/r_i))$. We show that the choice of the parameter $\mu > 0$ in this additional a priori assumption endows the Bayesian estimate of the regression dependence (13) with a certain level of selectivity in fusing the object-representation kernels. As a result, it will be enough to apply the Akaike Information Criterion only to the single selectivity parameter.

3.3   Estimation with Free Kernel-Related Variances and the Fixed Selectivity Level

Let the reciprocated positive variance coefficients $(1/r_1, \ldots, 1/r_n)$ be unknown and a priori considered as independently gamma distributed. In the standard definition, the family of gamma distributions $\gamma((1/r_i) \mid \alpha, \beta) = (\beta^\alpha / \Gamma(\alpha))(1/r_i)^{\alpha-1} \exp(-\beta(1/r_i))$ contains two parameters $\alpha > 1$ and $\beta > 0$ jointly determining the mathematical expectation $E(1/r_i) = \alpha/\beta$ and variance $Var(1/r_i) = \alpha/\beta^2$. We set $\alpha = (1/2)$ $[(1/\xi)(1 + 1/\mu) + 1]$, $\beta = 1/2\xi\mu$, and have now a parametric family of distributions defined only by $\mu \geq 0$, such that $E(1/r_i) = (1 + \xi)\mu + 1$ and $Var(1/r_i) = 2\xi\mu[(1 + \xi)\mu + 1]$. If $\mu \to 0$, a priori random values $1/r_i$ approach identity $1/r_i \cong \ldots \cong 1/r_n \cong 1$, however, if μ grows, the independent nonnegative values $1/r_i$ may differ arbitrarily, because $Var(1/r_i)$ increases much faster than $E(1/r_i)$.

The joint a priori distribution of independent inverse variances is proportional to the product

$$G(1/r_1, \ldots, 1/r_n \mid \mu, \xi) \propto \left(\prod_{i=1}^n 1/r_i\right)^{(1/2)[(1/\xi)(1+1/\mu)-1]} \exp\left(-(1/2\xi\mu)\sum_{i=1}^n (1/r_i)\right),$$

so, the Bayesian training criterion (10) with additional respect to the a priori distribution of variances will have the form

$$
\begin{aligned}
(\hat{c}_1, \ldots, \hat{c}_n, \hat{b}, \hat{r}_1, \ldots, \hat{r}_n) &= \arg\max P(c_1 \in \tilde{\Omega}_1, \ldots, c_n \in \tilde{\Omega}_n, b, r_1 \geq \varepsilon, \ldots, r_n \geq \varepsilon \mid \Omega^*, \mu, \xi) \\
&= \arg\max [\ln \Phi(y_1, \ldots, y_N \mid \omega_1, \ldots, \omega_N; a_1, \ldots, a_n, \xi) + \ln \Psi(c_1, \ldots, c_n, b \mid r, \ldots, r_n, \xi) \\
&\quad + \ln G(1/r_1, \ldots, 1/r_n \mid \mu, \xi)].
\end{aligned}
$$

(14)

where $\varepsilon > 0$ is a small real number close to zero. Just as in (10), the Bayesian estimate does not depend on ξ.

Theorem 2. The regression model (7) inferred from the training set Ω^* under the a priori assumption of fixed regression coefficient in accordance with the Bayesian condition (14) has the closed form (5) with parameters $(\hat{a}_{ij}, \hat{b})$, which are the minimum point of the quadratic optimization criterion

$$
\begin{aligned}
J(a_{ij}, r_i, i &= 1, \ldots, n, b, \delta_j, j = 1, \ldots, N \mid \mu) \\
&= \sum_{i=1}^n \left((1/r_i)\left(\sum_{j=1}^N \sum_{l=1}^N K_i(\omega_j, \omega_l) a_{ij} a_{il} + 1/\mu\right) + (1 + 1/\mu) \ln r_i\right) \\
&\quad + \sum_{j=1}^N \left(y_j - \sum_{i=1}^n \sum_{l=1}^N K_i(\omega_j, \omega_l) a_{il} - b\right)^2 \to \min.
\end{aligned}
$$

(15)

The minimization of this criterion is provided by the Gauss-Seidel iterations

$$\left((r_i)^0 = 1, i = 1,\ldots,n\right), k = 0,$$

$$\left((r_i)^k, i = 1,\ldots,n\right) \Rightarrow \left((\delta_j)^k, b^k, i = 1,\ldots,n, j = 1,\ldots,N\right), (a_{ij})^k = (r_i)^k (\delta_j)^k,$$
$$\left((a_{ij})^k = (r_i)^k (\delta_j)^k, i = 1,\ldots,n, j = 1,\ldots,N\right) \Rightarrow \left((r_i)^{k+1}, i = 1,\ldots,n\right),$$

implemented by, in turn, solving the system of linear equations

$$\begin{cases} \left(\sum_{i=1}^{n} (r_i)^k K_i(\omega_j, \omega_j) + 1\right) \delta_j + \sum_{l=1, l \neq j}^{N} \left(\sum_{i=1}^{n} (r_i)^k K_i(\omega_j, \omega_l)\right) \delta_l + b = y_j, j = 1,\ldots,N, \\ \sum_{j=1}^{N} \delta_j = 0, \end{cases} \quad (16)$$

and computing by independent formulas

$$(r_i)^{k+1} = \frac{\sum_{j=1}^{N} \sum_{l=1}^{N} K_i(\omega_j, \omega_l)(\delta_j)^k (\delta_l)^k + 1/\mu}{1 + 1/\mu}, \quad i = 1,\ldots,n \quad (17)$$

The stopping rule of the iterative process of learning can be defined, for example, on the condition of convergence of sequences $r_i, i = 1,\ldots,n : \frac{1}{n}\sum_{i=1}^{n}\left|r_i^{k+1} - r_i^k\right|$ $< \varepsilon, \varepsilon > 0$. Stopping at a step results in an approximation to the Bayesian estimate of the regression model (5)

$$\hat{y}(\omega \,|\, \Omega^*, \mu) = \sum_{j=1}^{N} \hat{\delta}_j \sum_{i=1}^{n} \hat{r}_i K_i(\omega_j, \omega) + \hat{b}. \quad (18)$$

Convergence of the procedure occurs in $10-15$ steps for typical problems, suppressing redundant kernels through the allocating of very small (but always non-zero weights) $\hat{r}_i$ in the regression model (18).

The level of kernel selectivity is parametrically determined by $\mu : 0 < \mu < \infty$. As $\mu \to 0$, variances tend toward unity $\left((r_i)^{k+1} = 1, i = 1,\ldots,n\right)$ (17), and the training rule (16)–(17) degenerates to the non-selective regression estimation (12). Contrarily, when $\mu \to \infty$, iteration of the kernel weights $(r_i)^{k+1} = (r_i)^k \sum_{j=1}^{N} \sum_{l=1}^{N} K_i(\omega_j, \omega_l)$ $(\delta_j)^k (\delta_l)^k$ (17) is extremely selective.

Given a training set Ω^*, each value of the selectivity parameter $0 < \mu < \infty$ produces a collection of kernel weights $(\hat{r}_1(\mu),\ldots,\hat{r}_n(\mu))$, i.e., a version of the model density $\Psi(c_1,\ldots,c_n, b \,|\, \hat{r}_1(\mu),\ldots,\hat{r}_n(\mu), \xi)$ (9) of the hidden regression coefficients in the observation density $\Phi(y_1,\ldots,y_N \,|\, \omega_1,\ldots,\omega_N; c_1,\ldots,c_n, \xi)$ (8).

4 Experimental Results

4.1 Simulation Studies

The data are simulated from the following linear model $\mathbf{y} = \mathbf{Xc} + \xi, \xi : N(0, \rho)$.

First of all we illustrate the modalities selection ability of proposed algorithm by simple ground-truth example. We considered the regression problem model with 49 features, 100 objects, and the target variable computed as $y = x_2 + 3x_6 + 2x_{22} + \xi$ where $x_i : N(x_i|0,1), \xi : N(\xi|0,0.5)$.

We received $\hat{r}_2 = 1.00008$, $\hat{r}_6 = 9.01475$, $\hat{r}_{22} = 3.958091$, $\hat{r}_i < 0.001524$, $\forall i \notin \{2,6,22\}$ and $\hat{c}_2 = 1.00006$, $\hat{c}_6 = 3.00379$, $\hat{c}_{22} = 3.958091$, $\hat{c}_i \cong 0, \forall i \notin \{2,6,22\}$. You can see that coefficients were close to the true values.

The purpose of next simulation is to show that the Supervised Selectivity Approach not only dominates the lasso and elastic net in terms of prediction accuracy, but also is a better variable selection procedure.

We simulate 100 data sets each of which has 1000 objects: 20 observations for the training set and 980 for the testing set. In simulations we set the number of features $n = 20, 100, 500$. The noise variance in simulated data is 10 % of the variance of observed variable. In Simulation 1 we set $c_i = (1,1,0,\ldots,0)$, The design matrix $\mathbf{X}$ is generated from the multivariate normal distribution with mean 0, variance 1. In Simulation 2 we set $c_{1,2} = 1$, $c_{2,\ldots,15} = (0.7)^{i-1}$, $c_{16,\ldots,n} = 0$ and leave other setups the same as in Simulation 1. In Simulation 3 we set $c_{1,2} = 1, c_{3,\ldots,n} = (0.75)^{i-1}$ for $n = 20$, $c_{1,2} = 1, c_{3,\ldots,n} = (0.95)^{i-1}$ for $n = 100$ and $c_{1,2} = 1, c_{3,\ldots,n} = (0.99)^{i-1}$ for $n = 100$. The structural parameters of all methods are chosen by cross-validation, leave-one-out procedure.

We compare the prediction accuracy of the three methods using the average of the prediction mean-squared errors MSE for observed variable and coefficients. The results are summarized in Tables 1, 2 and 3.

Table 1. Comparison of the three methods (Lasso, EN and Supervised Selectivity) on prediction and variable selection accuracy for Simulation 1

n	Lasso	Elastic Net	SS
Prediction MSE			
20	0.2359	0.2316	0.0887
100	0.2560	0.2528	0.1990
500	0.3115	0.3114	0.4150
Variable selection accuracy			
20	0.1225	0.1164	0.0287
100	0.2314	0.1332	0.0562
500	0.2778	0.2176	0.0476

4.2 Real Data Examples

The real data was received from the well-known UCI repository (http://archive.ics.uci.edu/ml). All datasets are parted into training and test set as 80:20. In Fig. 1 we show number of selected features on real data. The dark part of each column indicates the number of selected features. Average mean-squared errors over 100 experiments are summarized in Table 4.

Table 2. Comparison of the three methods (Lasso, EN and Supervised Selectivity) on prediction and variable selection accuracy for Simulation 2

n	Lasso	Elastic Net	SS
Prediction MSE			
20	0.3474	0.3535	0.1512
100	0.6578	0.6226	0.5639
500	0.8813	0.8878	0.9575
Variable selection accuracy			
20	0.1397	0.1298	0.0319
100	0.1861	0.1642	0.0483
500	0.3045	0.2567	0.0611

Table 3. Comparison of the three methods (Lasso, EN and Supervised Selectivity) on prediction and variable selection accuracy for Simulation 3

n	Lasso	Elastic Net	SS
Prediction MSE			
20	0.3887	0.3919	0.1777
100	0.9937	0.9730	0.9128
500	1.0243	1.0448	1.0814
Variable selection accuracy			
20	0.1598	0.1123	0.0331
100	0.1914	0.1594	0.0564
500	0.3117	0.2892	0.0679

Table 4. The root mean square deviation (*mean* ± *SD*) for different algorithms on real data

Data	Lasso	Elastic Net	SCS
Auto-mpg	3.47 ± 0.14	3.47 ± 0.14	3.46 ± 0.13
Boston	5.06 ± 0.23	5.07 ± 0.24	5.05 ± 0.23
Diabetes	55.31 ± 0.33	55.36 ± 0.32	55.17 ± 0.29
Prostate	0.81 ± 0.11	0.82 ± 0.11	0.80 ± 0.09

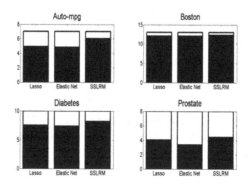

Fig. 1. Modalities selection on real data

5   Conclusions

We propose a new Bayesian approach to variable selection for dependency estimation. Real data examples and simulation studies show that Supervised Selectivity Shrinkage performs significantly better than EN and Lasso in prediction accuracy. Simulation studies suggest that SCS outperforms EN and Lasso in variable selection.

Acknowledgements. This research is funded by RFBR, grants 14-07-00964, 14-07-00527.

References

1. Vapnik, V.: Estimation of Dependencies Based on Empirical Data. Springer, New York (1982)
2. Schölkopf, B., Smola, A.J.: Learning with Kernels: Support Vector Machines, Regularization, Optimization, and Beyond. MIT Press, Cambridge (2001)
3. Bach, F.R., Lanckriet, G.R., Jordan, M.I.: Multiple kernel learning, conic duality, and the SMO algorithm. In: Proceedings of the Twenty-First International Conference on Machine Learning, p. 6. ACM (2004)
4. Fan, J., Lv, J.: A selective overview of variable selection in high dimensional feature space. Statistica Sinica **20**, 101–148 (2010)
5. Tatarchuk, A., Sulimova, V., Windridge, D., Mottl, V., Lange, M.: Supervised selective combining pattern recognition modalities and its application to signature verification by fusing on-line and off-line kernels. In: Benediktsson, J.A., Kittler, J., Roli, F. (eds.) MCS 2009. LNCS, vol. 5519, pp. 324–334. Springer, Heidelberg (2009)
6. Vapnik, V.: Statistical Learning Theory. Wiley, New York (1998)
7. Mottl, V.: Metric spaces admitting linear operations and inner product. Dokl. Math. **67**(1), 140–143 (2003)
8. Mottl, V., Krasotkina, O., Seredin, O., Muchnik, I.: Kernel fusion and feature selection in machine learning. In: Proceedings of the 8th IASTED International Conference on Intelligent Systems and Control, Cambridge, USA (2005)
9. Mottl, V., Krasotkina, O., Seredin, O., Muchnik, I.B.: Principles of multi-kernel data mining. In: Perner, P., Imiya, A. (eds.) MLDM 2005. LNCS (LNAI), vol. 3587, pp. 52–61. Springer, Heidelberg (2005)
10. Mottl, V.V., Seredin, O.S., Krasotkina, O.V., Muchnik, I.B.: Fusing of potential functions in reconstructing dependences from empirical data. Dokl. Math. **71**(2), 315–319 (2005). MAIK Nauka/Interperiodica
11. Mottl, V., Tatarchuk, A., Sulimova, V., Krasotkina, O., Seredin, O.: Combining pattern recognition modalities at the sensor level via kernel fusion. In: Haindl, M., Kittler, J., Roli, F. (eds.) MCS 2007. LNCS, vol. 4472, pp. 1–12. Springer, Heidelberg (2007)

Detecting Ordinal Class Structures

Raphael Lattke[2], Ludwig Lausser[1], Christoph Müssel[2,3],
and Hans A. Kestler[1,2(✉)]

[1] Leibniz Institute for Age Research – Fritz Lipmann Institute, 07745 Jena, Germany
[2] Core Unit Medical Systems Biology and Institute of Neural Information Processing,
Ulm University, 89069 Ulm, Germany
`hkestler@fli-leibniz.de, hans.kestler@uni-ulm.de`
[3] Institute of Number Theory and Probability Theory, Ulm University,
89069 Ulm, Germany

Abstract. Relying on an ordinal relationship among class labels, ordinal classifiers incorporate semantic knowledge about the classes into a purely data-driven multi-class classification task. Under the assumption that this relationship is reflected in feature space, these classifiers organize their internals according to this information. One essential step required is the identification of the true inter-class dependencies.

In this work, we now focus on the ability of cascaded ensemble classifiers to detect the relationships among ordinal classes. The minimal class sensitivity proves to be suitable to quantify this ability. This is an important problem, as for instance in medical applications often the true ordering of the classes is unknown or only partly known. We show that we can detect the ordinal class structure or its absence and that this ability depends on both the chosen base classifiers and the corresponding training schemes.

Keywords: Ordinal classification · Multi-class classification

1 Introduction

Classification is often considered as a purely data-driven approach in which classification models are adapted to a set of observations. This definition may be limiting if semantic knowledge about the class labels exist. In ordinal classification, an intrinsic ordering of the classes is assumed (e.g., small < medium < large) to be reflected in feature space. Even though conventional multi-class classification algorithms can be applied to such problems, ordinal classification algorithms can take advantage of the semantic knowledge on the ordering [8]. Incorrect class orderings should decrease their performance.

Many ordinal classification methods utilize ensemble techniques. These algorithms decompose an ordinal classification task into a system of binary ones and mainly differ in the employed fusion architecture. Frank and Hall propose

---

R. Lattke and L. Lausser—Contributed equally.

© Springer International Publishing Switzerland 2015
F. Schwenker et al. (Eds.): MCS 2015, LNCS 9132, pp. 100–111, 2015.
DOI: 10.1007/978-3-319-20248-8_9

an ordered linear sequence of binary logistic regression classifiers for estimating conditional class probabilities [7]. Cardoso and Pinto da Costa devise a cascaded ensemble classifier [5]. Hühn et al. utilize ordered binary trees for constructing ordinal ensemble classifiers [8]. Platt et al. represent the relations of base classifiers as a directed acyclic graph [11].

In this work, we assess whether ordinal ensemble classifiers can be utilized to identify the true ordering of a set of ordinal class labels. We investigate the influence of incorrect class orders on the classification performance of several ordinal cascaded ensemble classifiers in terms of the classifiers' minimal class sensitivities. Our experiments indicate that the tested models are able to reject incorrect class orders. Their performance depends both on the choice of the base classifiers and on the corresponding training scheme.

2   Methods

We will use the following notation throughout: a data object is represented by a vector $x = (x^{(1)}, ..., x^{(n)}) \in \mathcal{X}$ with $\mathcal{X}$ being the space of features. Each data object x_i is associated with a class label $y_i \in \mathcal{Y}$, with $\mathcal{Y}$ being a finite space of labels. A dataset of m data objects is defined as $S = \{(x_i, y_i)\}_{i=1}^m$. Classification denotes the prediction of a label $y_i \in \mathcal{Y}$ for a data object $x_i \in \mathcal{X}$. A suitable prediction model (the classifier) is determined from a training set $\mathcal{T} = \{(x_i, y_i)\}_{i=1}^m$ of samples for which the assignment of labels to data objects is known. This is done in an initial learning phase

$$l : \mathcal{T} \times \mathcal{C} \mapsto c_{\mathcal{T}}. \tag{1}$$

The model is selected from a concept class $c \in \mathcal{C}$ that characterizes the essential structural properties of a model type. After the learning phase, the model can be used to predict the label for an unseen data object by a mapping

$$c_{\mathcal{T}} : \mathcal{X} \longrightarrow \mathcal{Y}. \tag{2}$$

2.1   Multi-class Classification

Many classification algorithms (e.g. linear classifiers) are designed for binary classification tasks ($|\mathcal{Y}| = 2$). They are restricted to the classification of dichotomies and cannot be applied directly if there are more than two possible outcomes [3]. Various approaches have been proposed to extend binary classifiers to multi-class classifiers [9]. A common strategy is to decompose a multi-class classification task into a system of dichotomies that is handled by an ensemble of binary classifiers $\mathcal{E} = \{c_i : \mathcal{X} \mapsto \mathcal{Y}_i\}_{i=1}^{|\mathcal{E}|}$ [9,14]. Each of these classifiers is trained on a relabeled and possibly resampled version of the original dataset:

$$l : \mathcal{T}_{(y_+, \mathcal{Y}_+)} \cup \mathcal{T}_{(y_-, \mathcal{Y}_-)} \times \mathcal{C} \mapsto c_i, \tag{3}$$

where $y_\pm$ denotes the two new class labels and $\mathcal{Y}_\pm \subset \mathcal{Y}$, $\mathcal{Y}_+ \cap \mathcal{Y}_- = \emptyset$ denotes the two collections of classes that are merged into the new training sets

$$\mathcal{T}_{(y_\pm, \mathcal{Y}_\pm)} = \{(\boldsymbol{x}, y_\pm) | (\mathbf{x}, y) \in \mathcal{T}, y \in \mathcal{Y}_\pm\}. \tag{4}$$

The final prediction is based on information returned by all binary classifiers. In the following, we restrict ourselves to late fusion schemes, which rely on the predictions made by the base classifiers [13]. The corresponding ensemble classifier can be seen as a mapping

$$h_\mathcal{E} : \mathcal{Y}_1 \times ... \times \mathcal{Y}_{|\mathcal{E}|} \mapsto \mathcal{Y} \text{ with } h_\mathcal{E}(c_1(\boldsymbol{x}), ..., c_{|\mathcal{E}|}(\boldsymbol{x})) = y. \tag{5}$$

A common example for a multi-class ensemble is the one-against-one (OAO) fusion scheme [9]. It divides a k-class classification task into $|\mathcal{E}| = \frac{k(k-1)}{2}$ two-class tasks that are addressed independently by individual classifiers

$$c_{(y,y')} : \mathcal{X} \mapsto \{y, y'\} \text{ for all } y, y' \in \mathcal{Y}, y \prec y'. \tag{6}$$

Here $y \prec y'$ denotes an arbitrary total order among the class labels in $\mathcal{Y}$. The training of a single classifier is based on all samples of the corresponding classes

$$l : \mathcal{T}_{(y, \{y\})} \cup \mathcal{T}_{(y', \{y'\})} \times \mathcal{C} \mapsto c_{(y,y')}. \tag{7}$$

The OAO ensemble predicts the class label of an object according to a (fixed) majority vote

$$h_\mathcal{E}(c_1(\boldsymbol{x}), ..., c_{|\mathcal{E}|}(\boldsymbol{x})) = \underset{y \in \mathcal{Y}}{\operatorname{argmax}}(\sum_{c \in \mathcal{E}} \mathbb{I}_{[c(\boldsymbol{x})=y]}). \tag{8}$$

Classical multi-class ensembles are designed as purely data-driven algorithms. Knowledge on the semantics of classes and their interactions is neither taken into account for the design of a fusion architecture nor for the corresponding training algorithms.

2.2 Ordinal Classification

Ordinal classification is a special case of multi-class classification that relies on semantic knowledge on the class ordering [8]. Here, a total order of the class labels (e.g., small < medium < large) is known. We utilize class labels $\mathcal{Y} = \{1, ..., k\}$ in the following. It is assumed that the ordered structure of the label space is reflected in the topology (of a subspace) of the feature space. An example for such ordinal data is shown in Fig. 1. Here, a greater label is reflected by higher values of the shown features.

In this work we address the question whether the correct order of ordinal class labels can be identified via ordinal ensemble classifiers. More precisely, we analyze the performance of a cascaded fusion architecture [5,12] (Fig. 2). The cascaded classifier $h_\mathcal{E}$ uses an ensemble of $k - 1$ base classifiers

$$\mathcal{E} = \{c_{(i,i+1)} : \mathcal{X} \mapsto \{i, i+1\}\}_{i=1}^{k-1} \tag{9}$$

for solving the classification task. It evaluates the base classifiers iteratively. If the ith base classifier $c_{(i,i+1)}(\mathbf{x})$ predicts class i, the procedure stops, and the

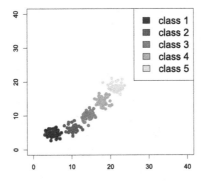

Fig. 1. Example of an ordinal classification task: Greater values for both features correspond to a greater class label. For complete description of the dataset, see Sect. 3.

ensemble classifier predicts class i. Otherwise the current sample is passed to the next base classifier. The ensemble classifier acts as a decision list and performs the mapping

$$h_{\mathcal{E}}(\mathbf{x}) = \min\left(|\mathcal{E}|, \underset{i \in 1,\ldots,|\mathcal{E}|-1}{\arg\min} [c_i(\mathbf{x}) = i]\right). \tag{10}$$

The cascaded fusion architecture itself can be seen as an untrainable late fusion architecture which is fixed for a given label order. Its performance only depends on the chosen type of base classifiers and their training. We utilize three different training schemes for base classifiers in our experiments:

Current vs. rest (CR): The ith classifier is trained to separate the ith class from those $\geq i+1$

$$l_{CR} : \mathcal{T}_{(i,\{i\})} \cup \mathcal{T}_{(i+1,\{i+1,\ldots,k\})} \times \mathcal{C} \mapsto c_{(i,i+1)} \tag{11}$$

Lower vs. higher (LH): The ith classifier is trained to separate all classes $\leq i$ from those $\geq i+1$

$$l_{LH} : \mathcal{T}_{(i,\{1,\ldots,i\})} \cup \mathcal{T}_{(i+1,\{i+1,\ldots,k\})} \times \mathcal{C} \mapsto c_{(i,i+1)}. \tag{12}$$

Pairwise (PW): The ith classifier is trained to separate the ith class from the $i+1$th class

$$l_{PW} : \mathcal{T}_{(i,\{i\})} \cup \mathcal{T}_{(i+1,\{i+1\})} \times \mathcal{C} \mapsto c_{(i,i+1)}. \tag{13}$$

3   Experiments

In these training schemes, all base classifiers are trained one by one. Each base classifier is adapted separately and it is not influenced by the training of the remaining ensemble. The base classifiers are especially not enforced to be parallel as e.g. in [5]. To answer the question whether cascaded classifiers can recover a

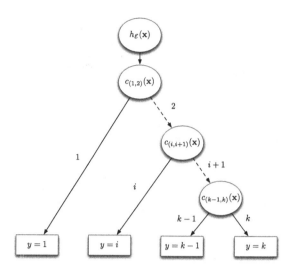

Fig. 2. Classification scheme of a cascaded classifier $h_{\mathcal{E}}(\mathbf{x})$. The base classifiers $c_{(i,i+1)}(\mathbf{x})$ are applied iteratively to a data object $\mathbf{x}$. If the ith classifier predicts class i, the procedure stops and $h_{\mathcal{E}}(\mathbf{x}) = i$. Otherwise, the next classifier in the cascade is applied to the data object.

potential ordinal structure of a dataset or not, we evaluate them on all possible $k!$ class orders of a k-class classification task. The cascaded classifiers are analyzed in $R \times F$ cross-validation experiments [3]. For a single F-fold cross-validation, the training dataset is split into F disjoint sets $\mathcal{P}_1, ..., \mathcal{P}_F$ of approximately equal size. The training of a model is performed on $\mathcal{T}_f = \bigcup_{i \in \{1,...,F\} \setminus \{f\}} \mathcal{P}_i$. This training set is decomposed for the training of the base classifiers of the cascaded classifier. The remaining set $\mathcal{P}_f$ is used to validate the model. The F-fold cross-validation is performed on R independent permutations of $\mathcal{T}$. All classifiers and training schemes are evaluated in 10×10 cross-validation experiments with identical splits.

The performance of a cascaded classifier is analyzed in terms of its class-wise sensitivities. In a $R \times F$ cross-validation, the sensitivity S_i of class i is estimated by

$$S_i = \frac{1}{R * F} \sum_{r=1}^{R} \sum_{f=1}^{F} \frac{1}{|\{(\boldsymbol{x}, y) \in \mathcal{P}_{r,f}, y = i\}|} \sum_{(\boldsymbol{x}, i) \in \mathcal{P}_{r,f}} \mathbb{I}_{[c_{\mathcal{T}_{r,f}}(\boldsymbol{x}) = i]}. \qquad (14)$$

We use the minimal class sensitivity

$$S_m = \min_{i \in \mathcal{Y}} S_i \qquad (15)$$

as a performance measure. A classifier that is not capable of capturing all classes appropriately will yield low values here.

Our analyses cover two permutation experiments: In the first experiment, we compare the minimal sensitivity of a cascaded classifier S_m to the minimal sensitivity S'_m of an OAO ensemble. The performance of a cascaded classifier is considered as comparable to that of an OAO ensemble if the corresponding $S_m >$ $0.9 \cdot S'_m$. In the second experiment, we address the question whether an ordinal cascaded classifier is able to recover the undirected ordinal structure of an ordinal classification problem. That is, we assess whether the minimal class sensitivity of the true class order (e.g. $1 < 2 < \ldots < k-1 < k$) and the corresponding reverse order (e.g. $k < k-1 < \ldots < 2 < 1$) can be distinguished from the minimal class sensitivities of any other pair of class order and corresponding reverse order. A pair of class orders is defined as detected if both orders achieve $S_m > 0.5$.

In our experiments, we utilize three different training schemes (CR, LH, PW) and three different types of base classifiers for the cascaded classifiers. As base classifiers, a linear support vector machine [15] (SVM), a 3-nearest neighbor classifier [6] (3NN) and a classification tree [4] (CART) are chosen. Reference experiments with non-ordinal multi-class classifiers are conducted either with an OAO ensemble or with a standard multi-class version of the base classifiers (MC) in case of 3NN and CART. All experiments were performed with the help of the TunePareto software [10].

An overview of the analyzed datasets is shown in Table 1. These datasets have been referred to as being truly ordinal (see references). Additionally, we employed an artificial dataset to examine basic properties of the cascaded classifier. This dataset comprises two features whose values increase with the class labels. Both features were exposed to random normally distributed noise. The dataset comprises five classes and is shown in Fig. 1.

Table 1. List of utilized datasets. The samples per class are provided in the order of the corresponding label space.

No.	Dataset	Features	Classes	Samples (per class)					
				Total	Class 1	Class 2	Class 3	Class 4	Class 5
AD	Artificial data	2	5	250	50	50	50	50	50
d_1	Balance scale [1]	4	3	625	288	49	288	-	-
d_2	cars [1]	6	4	1728	1210	384	69	65	-
d_3	CPU [8]	6	4	209	50	53	53	53	-
d_4	ESL [2]	4	5	488	52	100	116	135	85
d_5	journal [8]	5	3	172	91	53	28	-	-
d_6	LEV [2]	4	4	1000	93	280	403	224	
d_7	SWD [2]	10	3	1000	384	399	217	-	-

4   Results

The results on the artificial dataset AD are illustrated in Fig. 3. Here, all available samples (Fig. 1) are utilized for training the classifier ensembles. Panel Fig. 3(a) shows the decision regions of classifier ensembles that were supplied with the correct order of classes ($1 < 2 < 3 < 4 < 5$). In these experiments, the decision

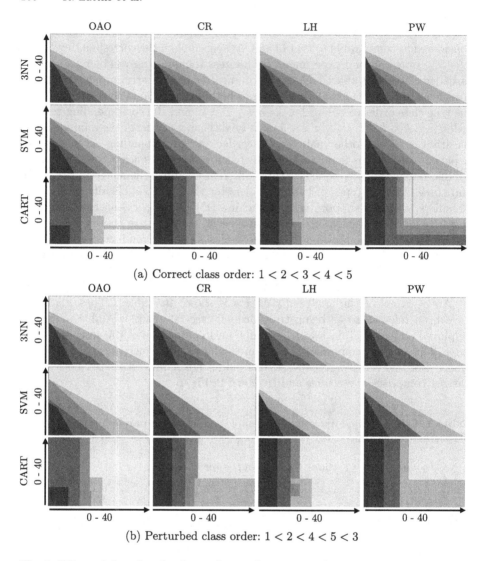

Fig. 3. Effects of changing the class order on the training of ordinal cascaded classifiers on the artificial dataset (Fig. 1). The decision regions for the correct order ($1 < 2 < 3 < 4 < 5$) are shown in Panel (a). The decision regions of a perturbed order ($1 < 2 < 4 < 5 < 3$) are shown in Panel (b). For the perturbed order, several cascades split the feature space only into four classes.

regions of 3NN ensembles and SVM ensembles are more similar to each other than to the decision regions of the CART ensembles. Among the SVM ensembles and the 3NN ensembles, the OAO training scheme and the CR training scheme generate more parallel decision boundaries than the CR training scheme and the PW training scheme. For the CART ensembles, the four training schemes

Table 2. Cross-validation errors of the standard multi-class versions (MC) and the OAO ensembles as compared to the cascaded classifier ensembles of the applied base classifiers. The MC version of the SVM is already implemented as OAO.

No.	3NN					SVM				CART				
	MC	OAO	CR	LH	PW	OAO	CR	LH	PW	MC	OAO	CR	LH	PW
AD	0.00	0.00	0.00	0.00	0.00	0.00	0.00	0.00	0.00	0.04	0.05	0.04	0.04	0.04
d_1	0.13	0.13	0.12	0.15	0.38	0.08	0.13	0.12	0.50	0.22	0.21	0.25	0.24	0.56
d_2	0.04	0.04	0.04	0.04	0.07	0.15	0.15	0.18	0.15	0.03	0.03	0.02	0.03	0.04
d_3	0.13	0.13	0.12	0.14	0.14	0.08	0.07	0.08	0.09	0.15	0.15	0.15	0.14	0.15
d_4	0.25	0.26	0.26	0.25	0.26	0.23	0.23	0.22	0.23	0.31	0.31	0.32	0.32	0.36
d_5	0.30	0.29	0.28	0.28	0.30	0.25	0.24	0.25	0.25	0.32	0.34	0.34	0.39	0.39
d_6	0.35	0.35	0.35	0.35	0.36	0.38	0.39	0.39	0.38	0.37	0.38	0.38	0.37	0.46
d_7	0.40	0.40	0.39	0.38	0.41	0.39	0.40	0.39	0.39	0.44	0.43	0.43	0.43	0.44

Table 3. Number of class orders for which the minimal class sensitivity S_m of the cascaded classifier is comparable to the minimal sensitivity S'_m of the corresponding OAO ensemble ($S_m \geq 0.9 \cdot S'_m$). A '*' indicates that OAO ensemble does not address the classification task properly ($S'_m < 0.5$). Experiments where only the true class order and the reverse order were found are highlighted.

No.	Total orders	3NN			SVM			CART		
		CR	LH	PW	CR	LH	PW	CR	LH	PW
AD	120	120	120	2	24	2	2	120	120	2
d_1	6	*	*	*	4	2	2	*	*	*
d_2	24	24	19	0	6	0	1	19	11	0
d_3	24	24	24	2	8	2	2	23	9	2
d_4	120	120	120	2	16	2	2	117	37	2
d_5	6	6	6	3	4	2	2	6	4	1
d_6	24	24	24	2	2	0	3	24	20	0
d_7	6	2	0	2	2	0	2	2	1	0

result in four different partitions of the feature space. The corresponding results of the 10×10 cross-validation error experiments are shown in Table 2. The 3NN ensembles as well as the SVM ensembles classify AD perfectly (0.00 CV error). The CART ensembles yield a maximal CV error of 0.05. Panel Fig. 3(b) shows the effects of providing an incorrect class order to the ensemble classifiers ($1 < 2 < 4 < 5 < 3$). The OAO architecture is mainly unaffected by this wrong information. Only the corresponding CART ensemble shows small differences to the ensemble trained on the correct class order. For the PW training scheme, all tested base classifiers divide the feature space into four instead of five classes. The CR training scheme and the LH training scheme show this behavior only for SVM ensembles. The corresponding 3NN ensembles produce a similar partitioning as

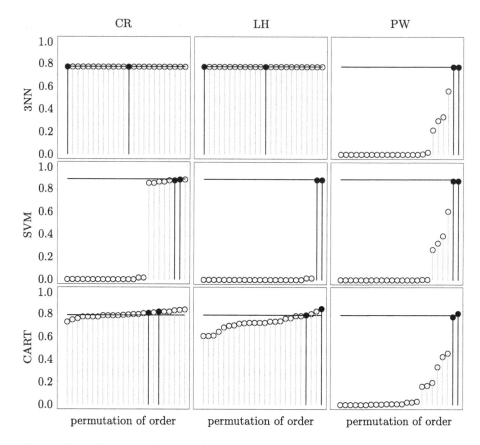

Fig. 4. Minimal class sensitivities of the cascaded classifiers for different label permutations. Sensitivities are sorted increasingly. The horizontal line indicates the performance of the OAO ensemble. The true order of the label space and the true reverse order are indicated by filled points.

for the correct class order. The decision regions of the CART cascades show different partitions for the correct and the incorrect class order.

All cross-validation errors are listed in Table 2. In general, the cascaded classifiers achieve cross-validation errors comparable to the corresponding MC and OAO architectures (all $p \geq 0.05$, two-sided Wilcoxon rank-sum tests and Holm correction). The CART ensembles show higher median cross-validation errors than the 3NN ensembles and SVM ensembles on 4 of 7 datasets (d_1, d_3, d_4, d_7). The SVM ensembles achieve a higher median cross-validation error on d_2. Among the cascaded classifiers, the lowest cross-validation errors are achieved either by the CR training scheme or by the LH training scheme. In combination with CART, the PW training scheme achieves the highest cross-validation error rates on six datasets (d_1, d_3, d_4, d_5, d_6, d_7). The PW training scheme also leads to a decreased performance on dataset d_2.

Table 4. Assessment of the ability of ordinal cascaded classifiers to recover the undirected class order based on the values of S_m for all possible order permutations. The undirected order is considered as recovered if $S_m > 0.5$ only for the true order and the true reverse order and for no other pair of order and reverse order. A '+' indicates that the true order could be recovered, a '−' indicates that this is not possible.

No.	3NN			SVM			CART		
	CR	LH	PW	CR	LH	PW	CR	LH	PW
AD	−	−	+	+	+	+	−	−	+
d_1	−	−	−	−	+	+	−	−	−
d_2	−	−	−	−	+	−	−	−	+
d_3	−	−	+	+	+	+	−	−	+
d_4	−	−	+	−	+	+	−	−	+
d_5	−	−	−	−	+	+	−	+	+
d_6	−	−	+	−	−	−	−	−	−
d_7	−	−	+	−	−	+	−	−	−

The effects of providing an incorrect class order to the cascaded classifier are analyzed in terms of S_m. The cross-validation experiments were repeated for all possible permutations of class orders. Figure 4 shows the results for the CPU dataset (d_3). For the true order ($1 < 2 < 3 < 4$) and its reverse order ($4 > 3 > 2 > 1$) all cascaded classifiers show high S_m values regardless of the chosen base classifier or training scheme. For all experiments with the PW training scheme, all other (incorrect) class orders exhibit considerably lower S_m. For both the CR and the LH training scheme, 3NN ensembles and CART ensembles achieve $S_m > 0.6$ for all incorrect class orders. This does not hold for SVM ensembles: Here, there is an obvious dichotomy between orders that achieve high values of S_m and orders that achieve very low values of S_m.

A summary of the first permutation experiment for all datasets can be found in Table 3. Here, the number of class orders for which the minimal class sensitivity S_m of a cascaded classifier is comparable to the minimal class sensitivity S'_m of the corresponding OAO ensemble ($S_m \geq 0.9 \cdot S'_m$) is shown. The response to a permuted class order strongly depends on both the training scheme of the cascaded classifier and the employed base classifier. Among the tested training schemes, the PW algorithm shows the highest susceptibility to changes of the class order. The susceptibility of both the LH training scheme and the CR training scheme depends on the base classifier that is used. For the SVM ensembles, these training schemes show high susceptibility to varying order. This effect cannot be observed for 3NN ensembles and CART ensembles. Among the base classifiers, the SVM algorithm generally shows the highest susceptibility to varying order, whereas the CART and the 3NN algorithm are less susceptible. Interestingly, in the experiments where only two class orders were considered as comparable to the OAO ensemble, the true class order and its reverse were selected in most cases. This means that the true class order (without considering the direction) was recovered. These cases are highlighted in Table 3.

Table 4 shows the results of the second permutation experiment. Here, a pair of true class order and reverse class order is considered as recovered if $S_m > 0.5$ for both orders and for no other perturbed pair of a class order and the corresponding reverse. The PW training scheme shows the highest power of recovering the true order. In combination with the SVM base classifier, it correctly identifies the true undirected class order for 6 of 8 datasets. For the 3NN ensembles and CART ensembles, it recovers the true undirected class order for 5 of 8 datasets. For the LH training scheme, the true order is returned in 6 of 8 cases when using SVMs with a linear kernel for the base classifiers. The CR training scheme did not identify the correct undirected class order in any experiment.

5   Conclusion

Our experiments address the question whether ordinal classification algorithms can be used for detecting the true order of ordinal classes. These learning algorithms rely the a priori information on the class order and assume this semantic knowledge to be reflected in the feature space. Our experiments with cascaded ensemble architectures indicate that ordinal classifiers exist whose performance is strongly influenced by the provided class orders. The strength of this influence depends on the chosen training scheme and the chosen type of base classifier. In our experiments, the best results were achieved for a pairwise training scheme in which each base classifier is trained on the samples of two neighbouring classes. When training each base classifier on all samples, only linear support vector machines can detect the correct class order. Training a classifier to separate the current class from all higher classes was least susceptible to an incorrect class order. Out of the tested base classifiers, linear support vector machines are most susceptible to the true class order.

All in all, cascaded classifiers based on linear support vector machines and a pairwise training scheme could eliminate the highest amount of perturbed class orders. Interestingly, the true class order and its reverse could not be distinguished in most of the experiments. Both these orders achieved comparable minimal class sensitivities. This might be a hint that the correctness of the neighbour relationship is more important than the direction of the global class order.

Acknowledgements. The research leading to these results has received funding from the European Community's Seventh Framework Programme (FP7/20072013) under grant agreement n°602783 (to HAK), the German Research Foundation (DFG, SFB 1074 project Z1 to HAK), and the Federal Ministry of Education and Research (BMBF, Gerontosys II, Forschungskern SyStaR, project ID 0315894A to HAK).

References

1. Bache, K., Lichman, M.: UCI machine learning repository (2013). http://archive. ics.uci.edu/ml
2. Ben David, A.: Dataset: Ordinal real-world datasets (esl, lev, swd). Irvine, CA.: University of California, School of Information and Computer Science (2008). http://archive.ics.uci.edu/ml
3. Bishop, C.: Pattern Recognition and Machine Learning. Springer, New York (2006)
4. Breiman, L., Friedman, J.H., Olshen, R.A., Stone, C.J.: Classification and Regression Trees. Wadsworth Publishing Company, Belmont (1984)
5. Cardoso, J., Pinto da Costa, J.: Learning to classify ordinal data: the data replication method. J. Mach. Learn. Res. **8**, 1393–1429 (2007)
6. Fix, E., Hodges, J.L.: Discriminatory analysis: Nonparametric discrimination: Consistency properties. Technical report. Project 21-49-004, Report Number 4, USAF School of Aviation Medicine, Randolf Field, Texas (1951)
7. Frank, E., Hall, M.: A simple approach to ordinal classification. In: Flach, P.A., De Raedt, L. (eds.) ECML 2001. LNCS (LNAI), vol. 2167, pp. 145–156. Springer, Heidelberg (2001)
8. Hühn, J., Hüllermeier, E.: Is an ordinal class structure useful in classifier learning? J. Data Min. Model. Manage. **1**(1), 45–67 (2009)
9. Lorena, A., de Carvalho, A., Gama, J.: A review on the combination of binary classifiers in multiclass problems. Artif. Intell. Rev. **30**, 19–37 (2008)
10. Müssel, C., Lausser, L., Maucher, M., Kestler, H.A.: Multi-objective parameter selection for classifiers. J. Stat. Softw. **46**(5), 1–27 (2012)
11. Platt, J.C., Shawe-Taylor, J., Cristianini, N.: Large margin dag's for multiclass classification. In: Proceedings of Advances in Neural Information Processing Systems 12, January 1999
12. Rivest, R.L.: Learning decision lists. Mach. Learn. **2**(3), 229–246 (1987)
13. Sammut, C., Webb, G.: Encyclopedia of Machine Learning. Springer, New York (2010)
14. Tax, D., Duin, R.: Using two-class classifiers for multiclass classification. In: International Conference on Pattern Recognition (2002)
15. Vapnik, V.: Statistical Learning Theory. Wiley, New York (1998)

Calibrating AdaBoost for Asymmetric Learning

Nikolaos Nikolaou[(✉)] and Gavin Brown

School of Computer Science, University of Manchester, Kilburn Building,
Oxford Road, Manchester M13 9PL, UK
{nikolaos.nikolaou,gavin.brown}@manchester.ac.uk

Abstract. *Asymmetric classification* problems are characterized by class imbalance or unequal costs for different types of misclassifications. One of the main cited weaknesses of *AdaBoost* is its *perceived* inability to handle asymmetric problems. As a result, a multitude of asymmetric versions of AdaBoost have been proposed, mainly as heuristic modifications to the original algorithm. In this paper we challenge this approach and propose instead handling asymmetric tasks by properly *calibrating* the scores of the original AdaBoost so that they correspond to probability estimates. We then account for the asymmetry using classic decision theoretic approaches. Empirical comparisons of this approach against the most representative asymmetric Adaboost variants show that it compares favorably. Moreover, it retains the theoretical guarantees of the original AdaBoost and it can easily be adjusted to account for changes in class imbalance or costs without need for retraining.

Keywords: Boosting · Cost-sensitive · Class imbalance · Classifier calibration

1 Introduction

Most real world classification problems are *asymmetric*. This asymmetry means that either the classes have different *prior probabilities* or the *costs* of different types of misclassifications are unequal, or both. A doctor testing a patient for a life-threatening disease, is faced with a cost-sensitive decision: a false positive will lead to further tests which will eventually reveal the misdiagnosis, while a false negative can be lethal. An astrophysicist predicting whether a telescope image contains a supernova or not faces an imbalanced class problem, as supernovae are rare.

AdaBoost [4] is a powerful, popular and recognized meta-learning technique. However it is often regarded as *skew-insensitive* [15,17], meaning it is unable to handle asymmetric tasks. There exist many *skew-sensitive* AdaBoost variants, including *AdaCost* [2,17], *CSB0, CSB1, CSB2* [17], *Asymmetric-Adaboost* [18], *RareBoost* [6], *AdaC1, AdaC2, AdaC3* [16], *CS-AdaBoost* [9,10]. However, most of them are heuristic and as a result they lack the theoretical guarantees of the original AdaBoost [7].

© Springer International Publishing Switzerland 2015
F. Schwenker et al. (Eds.): MCS 2015, LNCS 9132, pp. 112–124, 2015.
DOI: 10.1007/978-3-319-20248-8_10

It is also unclear if we really *need* to modify AdaBoost, or if asymmetric problems are better tackled by *calibrating* the *scores* of the original AdaBoost. Calibrated scores can be treated as *probability estimates* and can thus be used to handle the asymmetric nature of the task following *decision theoretic* approaches, similar to the classical work by Elkan [1]. For one, this approach can easily be adjusted to different class and cost imbalance setups, without the need to retrain the ensemble. Another benefit is that all favorable theoretical properties of the original AdaBoost will be preserved. The goal of this work is thus to investigate whether we can achieve comparable results to the asymmetric AdaBoost variants by calibrating the scores produced by the original AdaBoost and then choosing an appropriate classification threshold.

2   Background

2.1   Asymmetric Learning

In this paper we will be examining binary classification asymmetric problems, where an example can be either positive, denoted by a label $y = 1$ or negative, denoted by $y = -1$. The class imbalance can be captured by the different priors, $p(y = -1)$ and $p(y = 1)$, while the cost imbalance can be modeled with a *cost matrix* of the form

$$C = \begin{bmatrix} 0 & c \\ 1 & 0 \end{bmatrix}, \tag{1}$$

where 1 is the cost of a *false positive* and c the cost of a *false negative*[1]. The above matrix assigns a zero cost to all correct classifications, as is commonly the case [1].

Although *skewed class* and *skewed cost* problems are different [8], they can be formulated and treated in a similar way, by using a *skew ratio* c, that captures the *relative importance* of positives w.r.t. negatives to adjust for either [3]. We commonly assume w.l.o.g. that the important class (the 'rare' one in an imbalanced class scenario, or the 'expensive to misclassify' in a cost-sensitive scenario) is the positive one, $y = 1$. One difference is the evaluation measures used in each type of asymmetric problem. When facing a skewed cost task, the main goal is to minimize the total cost. When faced with a skewed class problem, the goal could instead be to achieve good performance on all classes.

Our analysis will focus on cost-sensitive tasks under a cost matrix of the form C, with skew ratio $c = c_{FN}/c_{FP} \geq 1$. This means that the cost of misclassifying the i-th example is

$$c(y_i) = \begin{cases} c, & \text{if } y_i = 1 \\ 1, & \text{if } y_i = -1 \end{cases},$$

---

[1] A more intuitive equivalent form is $\begin{bmatrix} 0 & c_{FN} \\ c_{FP} & 0 \end{bmatrix}$. Scaling the cost matrix has no effect on the decision problem, so we can divide its entries with c_{FP}, thus assigning a cost of 1 to false positives and a cost of $c = c_{FN}/c_{FP}$ to false negatives.

where y_i is the label of the instance. The primary evaluation measure we will use in this paper is the *average cost* incurred on the test set

$$L_{Avg}(\mathbf{x}'_i, y_i; H, c) = \frac{1}{N_{test}} \sum_{i=1}^{N_{test}} \mathbb{I}[H(\mathbf{x}'_i) \neq y_i] c(y_i),$$

where $H(\mathbf{x}'_i)$ is the prediction of the ensemble on the test example $\mathbf{x}'_i$ and $\mathbb{I}[\cdot]$ is the *indicator function* which outputs '1' iff its argument is true. We will also use *precision*, i.e. the fraction of positive predictions that are correct and *recall*, the fraction of positives correctly classified, to shed light on the different behaviors of the methods we examine. Precision and recall are given by

$$Prec = \frac{TP}{TP + FP}, \quad Rec = \frac{TP}{TP + FN},$$

where TP, FP and FN are the numbers of *true positives*, *false positives* and *false negatives* on the test set, respectively. Both quantities need to be high for prediction to be reliable. The harmonic mean of precision and recall,

$$F - measure = \frac{2 \cdot Rec \cdot Prec}{Rec + Prec},$$

can be used to capture both precision and recall at once. A high *F-measure* is desirable, as it implies that the values of both precision and recall are high.

2.2   AdaBoost

AdaBoost [4] is an *ensemble learning technique* which constructs a *strong classifier* H sequentially by combining multiple *weak classifiers* h_t, $t = 1, \ldots, M$. A weak classifier is one that is marginally more accurate than random guessing and a strong classifier is one that achieves arbitrarily high accuracy. AdaBoost achieves this by training each subsequent model h_t on a new dataset in which the examples misclassified by the previous model are assigned more weight and the ones that were correctly classified are assigned less weight. This can be achieved either by reweighing or by resampling the dataset on each round. This work uses the reweighing approach and focuses on AdaBoost with *confidence rated predictions* [14], where each *base learner* h_t is assigned a *confidence score* α_t.

The algorithm is given as input a set of training examples of the form $(\mathbf{x}_i, y_i)$, $i = 1, \ldots, N$ where $\mathbf{x}_i$ is the feature vector of the i-th example and y_i is its class label. On the first round of AdaBoost, all training examples are assigned equal weights $D_i^1 = \frac{1}{N}$. On each round $t = 1, \ldots, M$, the weak learner h_t that minimizes the misclassification error $\epsilon_t = \sum_{i:h_t(\mathbf{x}_i) \neq y_i} D_i^t$, where $h_t(\mathbf{x}_i)$ is the predicted class of the i-th example by the t-th weak learner, is added to the ensemble. The confidence of weak learner h_t is computed as

$$\alpha_t = \frac{1}{2} \log \left(\frac{1 - \epsilon_t}{\epsilon_t} \right). \tag{2}$$

The weight of each example $i = 1, \ldots, N$ is then updated to

$$D_i^{t+1} = e^{-y_i h_t(\mathbf{x}_i)\alpha_t} D_i^t \tag{3}$$

and renormalized by $D_i^{t+1} \leftarrow \frac{D_i^{t+1}}{\sum_{i=1}^N D_i^{t+1}}$ so that $\sum_{i=1}^N D_i^{t+1} = 1$. These will be the weights of each example on the next round. The algorithm terminates when the maximum number M of weak learners have been added to the ensemble or when a base learner h_t with $\epsilon_t < 1/2$ cannot be found[2]. The *final prediction* on a test datapoint $\mathbf{x}'$ is given by the sign of the weighted sum of the weak learner predictions $h_t(\mathbf{x}')$ *weighted* by their corresponding confidence scores

$$H(\mathbf{x}') = sign\left[\sum_{t=1}^M \alpha_t h_t(\mathbf{x}')\right]. \tag{4}$$

2.3   Classifier Calibration

Many classifiers can have their output normalized to return a *score* $s(\mathbf{x}') \in [0, 1]$ for each test example $\mathbf{x}'$ indicating 'how positive' it is. In the case of AdaBoost, this score is the quantity $s(\mathbf{x}') = \frac{\sum_{t=1}^M \alpha_t \frac{h_t(\mathbf{x}')+1}{2}}{\sum_{t=1}^M \alpha_t}$. However, when a cost-sensitive decision needs to be made on the instance $\mathbf{x}'$, the score $s(\mathbf{x}')$ is of little use. Instead, we need to estimate the probability of $\mathbf{x}'$ belonging to the positive class $\hat{p}(y = 1|\mathbf{x}')$. This will allow us to assign $\mathbf{x}'$ to the class that *minimizes the expected cost*. In other words, in binary classification, $\mathbf{x}'$ is assigned to the positive class only if

$$\hat{p}(y = 1|\mathbf{x}')c > \hat{p}(y = -1|\mathbf{x}') \iff \hat{p}(y = 1|\mathbf{x}') > \frac{1}{1+c},$$

under the cost matrix of Eq. (1), making use of $\hat{p}(y = -1|\mathbf{x}') = 1 - \hat{p}(y = 1|\mathbf{x}')$. Otherwise, $\mathbf{x}'$ is assigned to the negative class.

The procedure of converting classifier scores to actual probability estimates is called *calibration*. A classifier is calibrated if $\hat{p}(y = 1|\mathbf{x}') \rightarrow s(\mathbf{x}')$, as $N \rightarrow \infty$, for any $\mathbf{x}'$ [21]. However, it has been previously noted [11] that as the number of boosting rounds increases, the scores $s(\mathbf{x}')$ get more pushed away from 0 or 1, exhibiting an increasing "sigmoid distortion". In other words, the scores produced by AdaBoost are a sigmoid transformation of actual probability estimates. A theoretical justification for this effect is based on the statistical interpretation of AdaBoost by Friedman et al. [5], under which AdaBoost is a *stagewise* procedure of constructing an *additive logistic regression model* which finds the weak learners h_t and their corresponding confidence scores α_t that minimize the *average exponential loss* across all training examples.

The results of Niculescu-Mizil and Caruana [11] showed empirically that once properly calibrated, AdaBoost produced better probability estimates than any other model examined. The authors corrected for the "sigmoid distortion" of

---

[2] Note that in the binary classification case, a hypothesis h_t with error $\epsilon_t > 1/2$ can be turned into one with $\epsilon_t < 1/2$ simply by flipping its predictions.

the AdaBoost scores using three different approaches. The first approach was to directly apply a *logistic correction* implied by the framework of Friedman et al. [5]. The second calibration method was *Platt scaling* [12], originally used to map *SVM* outputs[3] to posterior probabilities. Platt scaling consists of finding the parameters A and B for a sigmoid mapping $\hat{p}(y = 1|\mathbf{x}') = \frac{1}{1+e^{As(\mathbf{x}')+B}}$, such that the likelihood of the data is maximized. Fitting the parameters A and B requires the use of a separate validation set. Finally, they also performed calibration using *isotonic regression* [13]. The latter is non-parametric and more general as it can be used to calibrate scores which exhibit any form of monotonic distortion [20]. Platt scaling produced the most reliable probability estimates on small sample sizes among the three methods, closely followed by isotonic regression. In this paper we will therefore be calibrating the scores of AdaBoost using Platt scaling. The *Calibrated AdaBoost* algorithm is given in Fig. 1.

3   Asymmetric Boosting Algorithms

In the introduction we mentioned a number of AdaBoost variants proposed to handle asymmetric learning tasks. Most of these methods are proposed heuristically, i.e. by introducing ad-hoc changes to steps of the AdaBoost algorithm, rather than by starting from a cost-sensitive problem formulation, e.g. by defining a different *loss function* in place of AdaBoost's exponential loss. Asymmetric boosting methods can broadly be classified into two groups: those that modify the prediction rule Eq. (4) of AdaBoost and those that introduce modifications in the training phase, either by modifying the weight update rule of Eq. (3), or the calculation of the α_t coefficients of Eq. (2).

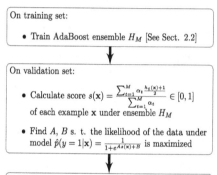

On training set:
- Train AdaBoost ensemble H_M [See Sect. 2.2]

On validation set:
- Calculate score $s(\mathbf{x}) = \frac{\sum_{t=1}^{M} \alpha_t \frac{h_t(\mathbf{x})+1}{2}}{\sum_{t=1}^{M} \alpha_t} \in [0,1]$ of each example $\mathbf{x}$ under ensemble H_M
- Find A, B s. t. the likelihood of the data under model $\hat{p}(y = 1|\mathbf{x}) = \frac{1}{1+e^{As(\mathbf{x})+B}}$ is maximized

On test set:
- Calculate score $s(\mathbf{x})$, $\forall$ example $\mathbf{x}$ under H_M
- Apply transformation $\hat{p}(y = 1|\mathbf{x}) = \frac{1}{1+e^{As(\mathbf{x})+B}}$ to the scores $s(\mathbf{x})$ to get probability estimates
- Predict class $H_M(\mathbf{x}) = sign[\hat{p}(y = 1|\mathbf{x}) - \frac{1}{1+c}]$

Fig. 1. Calibrated AdaBoost

3.1   Methods that Modify the Prediction Rule

A straightforward way to make AdaBoost skew-sensitive is to substitute the weighted majority vote prediction rule of Eq. (4) with the *minimum expected cost (MEC)* prediction rule

$$H_M(\mathbf{x}') = sign\left[\sum_{y\in\{-1,1\}} c(y)\sum_{t=1}^{M}\alpha_t h_t(\mathbf{x}')\right], \qquad (5)$$

---

[3] The mapping of outputs of SVMs to posterior probability estimates exhibits a similar sigmoid distortion to that observed in AdaBoost.

which reduces to Eq. (4) for $c = 1$, i.e. when the costs of false positives and false negatives are equal. This method trains a standard AdaBoost ensemble and only changes the decision rule used for the final prediction. The idea is to assign $\mathbf{x}'$ to the class that minimizes the expected cost. This approach has been mentioned in [17] without being given a specific name. In this paper we refer to it as *AdaMEC*.

3.2   Methods that Modify the Training Algorithm

CSB2 [17], changes the weight update rule of the original AdaBoost, given in Eq. (3), to

$$D_i^{t+1} = e^{-y_i h_t(\mathbf{x}_i)\alpha_t} C_{\delta(i)} D_i^t, \qquad where \quad C_{\delta(i)} = \begin{cases} 1, & \text{if } h_t(\mathbf{x}_i) = y_i \\ c(y_i), & \text{if } h_t(\mathbf{x}_i) \neq y_i. \end{cases} \qquad (6)$$

The form of the update rule of *CSB2* is the same as that of the original AdaBoost only for correctly classified examples, hence true positives and true negatives are not treated differently. On the other hand, misclassified examples have their weight updates adjusted by an multiplicative cost factor $c(y_i)$, thus false positives and false negatives are treated differently. *CSB2* reduces to AdaBoost for $c = 1$.

   AdaC2 [16] substitutes the weight update rule of the original AdaBoost, given in Eq. (3), by

$$D_i^{t+1} = e^{-y_i h_t(\mathbf{x}_i)\alpha_t} c(y_i) D_i^t, \qquad (7)$$

which also treats true positives and true negatives differently, unlike Eq. (6). The method also modifies the calculation of the α_t coefficients of Eq. (2) to

$$\alpha_t = \frac{1}{2} \log \frac{\sum_{i:h_t(\mathbf{x}_i)=y_i} D_i^t c(y_i)}{\sum_{i:h_t(\mathbf{x}_i)\neq y_i} D_i^t c(y_i)}. \qquad (8)$$

It is worth noting that *AdaC2* can be justified theoretically [15] as a stage-wise minimization of a *cost-weighted* version of the exponential loss, which for a classifier h_t, on an example $(\mathbf{x}_i, y_i)$ has the form $L(h_t(\mathbf{x}_i), y_i) = c(y_i)e^{-y_i h_t(\mathbf{x}_i)}$. Under this definition, the α_t calculated by Eq. (8) is *optimal*. Like the other two variants we described, when $c = 1$, *AdaC2* reduces to AdaBoost.

4   Empirical Evaluation

4.1   Experimental Setup

In our experiments we compare the performance of *AdaMEC*, *CSB2* and *AdaC2* to that of the *original AdaBoost calibrated with Platt scaling* under various degrees of cost skew, namely $c \in \{1, 1.5, 2, 2.5, 5, 10\}$. As a primary measure of performance we use the average cost attained on the test set. We also provide precision, recall and F-measure results to better demonstrate the different behavior of each method. As a base learner, we used univariate logistic regression trained with batch gradient descent. The maximum number of base learners M was set to 100.

We used 7 datasets from the *UCI repository*. Any entries with missing values were discarded. Our goal is to investigate the performance of each approach under various degrees of cost skew c. The datasets are originally imbalanced, so we selected an equal number of positive and negative examples, to suppress the additional effects of class imbalance. This was achieved by uniformly undersampling the majority class rather than by oversampling the minority class, as it avoids overfitting due to occurrences of identical examples in training, testing and validation sets. A summary of the datasets is given in Table 1.

We use a random 25 % of the data for testing. The remaining 75 % was used for training. In the case of calibration using Platt scaling, we needed to also reserve a separate validation set to fit the parameters of the sigmoid without overfitting. A third of the training data was used to this end. After training the models and –where applicable– calibrating on the validation set, we evaluated them on the test set. The entire procedure is repeated 30 times. For each method and evaluation measure, we report average values across all 30 runs as well as 95 % confidence intervals.

Table 1. Characteristics of the datasets used in this study. The table indicates the number of instances used, the number of features, and the class we chose to be 'positive' according to the naming convention in the original file. For example, in *semeion*, class '1' was chosen as 'positive' and the rest grouped under the 'negative' label.

Dataset	# Instances	Positive class	Negative class	# Features
survival	162	2	1	3
liver	290	1	2	6
pima	576	1	0	8
heart	240	1	0	13
wdbc	424	1	0	31
sonar	194	0	2	60
semeion	322	1	$\{2, ..., 10\}$	256

4.2   Analysis of Experimental Results

Average Cost: In terms of average cost, we observe different trends on the lower-dimensional datasets *survival, liver* and *pima* and on the higher-dimensional datasets, *wdbc, heart, semeion* and *sonar*. Results for *liver, pima* and *sonar* are omitted due to lack of space. When the problem is cost-insensitive ($c = 1$), all methods exhibit more or less the same performance. The performance of most methods also tends to be equivalent when the cost ratio is very high ($c = 10$), since for such high degrees of imbalance all examples tend to be assigned to the positive class. Our results are summarized in Fig. 2.

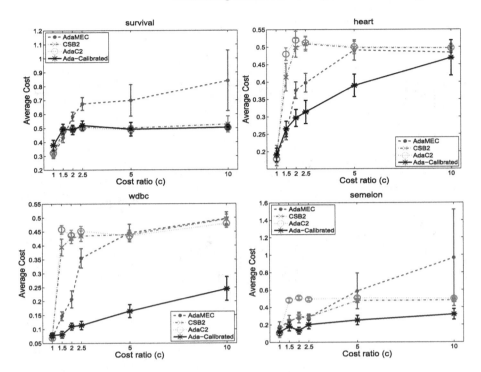

Fig. 2. Average cost results under various degrees of cost imbalance c. The cost attained by *calibrated AdaBoost* is lower than that of *AdaMEC*, *CSB2* and *AdaC2* on higher dimensional datasets like *heart*, *wdbc* and *semeion* and comparable on lower dimensional datasets like *survival*.

On low-dimensional datasets, the performance of *calibrated AdaBoost* is on par with that of *CSB2* and *AdaC2* and all three methods clearly outperform *AdaMEC*, which exhibits a high variance as c increases. This can partly be explained by the fact that *AdaMEC* used on average a much smaller number of weak learners than *CSB2* and *AdaC2*. On the other hand, *calibrated AdaBoost* tended to slightly fewer weak learners than *AdaMEC*. So its improved performance over *AdaMEC* can only be attributed to the calculation of more reliable probability estimates.

On higher-dimensional datasets, the performance of *calibrated AdaBoost* is even more impressive, as it clearly outperforms all other methods. *AdaMEC* exhibits the second-best performance at low degrees of skew c. *CSB2* and *AdaC2* produce the highest average cost with the former producing marginally lower average cost than the latter for low values of c.

Precision and Recall: The precision and recall curves reveal more details about the different behaviour of each method. *Calibrated AdaBoost* has the overall highest precision scores, even for high degrees of cost skew c. *AdaMEC* achieves the second-best overall precision. *CSB2* typically gives poor precision

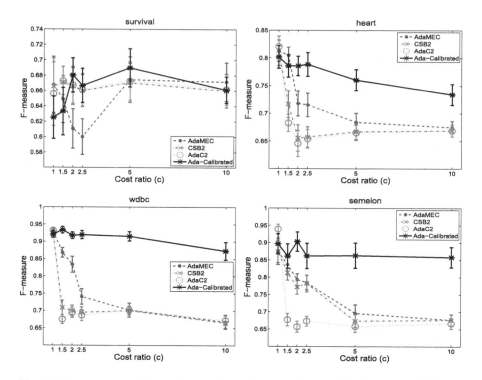

Fig. 3. F-measure results under various degrees of cost imbalance *c*. *Calibrated AdaBoost* produces higher F-measure scores than its competitors on the higher dimensional datasets *heart*, *wdbc* and *semeion* and comparable on the lower dimensional datasets like *survival*. It also shows a remarkable robustness to changes in *c*.

values, but on *semeion, heart* and marginally on *wdbc*, it still outperforms *AdaC2* for low values of *c*. We can again notice the two different trends regarding the *calibrated AdaBoost*. It does not outperform its competitors on the low-dimensional datasets *survival, pima* and *liver*, but its performance is comparable to theirs.

In terms of recall, all methods exhibit very high scores, close to the maximal value of 1. This is indicative of the cost-sensitive methods' eagerness to assign test instances to the positive class. *AdaC2* and *CSB2* have the overall highest recall, reaching the value of 1 even for small values of *c*. *AdaMEC* has the second-highest overall scores and *calibrated AdaBoost* exhibits the lowest recall values among the compared methods. This does not mean that *calibrated AdaBoost* behaves poorly in terms of recall, just that it is less *aggressive* than its competitors.

These results indicate that *AdaC2* is the most aggressive among the compared methods, as it tends to assign all test examples to the positive class even for relatively low values of *c*. This leads to zero false negatives, hence maximal recall, but also to many false positives, hence low precision. This behaviour of *AdaC2* is largely mimicked by *CSB2*. The next most aggressive method is *AdaMEC* and the least aggressive is calibrated AdaBoost. The results are summarized in (Fig. 4).

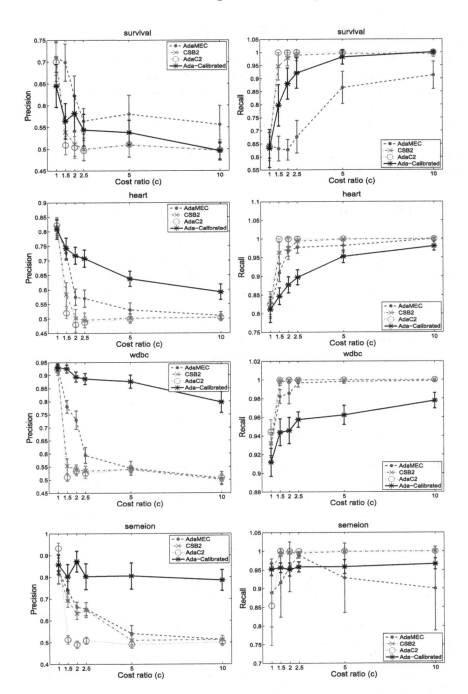

Fig. 4. Precision and recall results under various degrees of cost imbalance c. *Calibrated AdaBoost* achieves higher precision and lower recall than its competitors, especially on the higher dimensional datasets *heart*, *wdbc* and *semeion*.

F-Measure: On the F-measure curves of Fig. 3, we can again observe that on the low-dimensional datasets *survival, pima* and *liver* all methods exhibit comparable performance. This is not surprising, as problems with small numbers of features are generally easier than high-dimensional ones. Where the *calibrated AdaBoost* shines is on the higher dimensional datasets *wdbc, heart, semeion* and *sonar*. On these datasets it attains F-measure values far higher than the asymmetric AdaBoost variants, exhibiting an admirably small sensitivity to the cost ratio c. The F-measure values of *AdaMEC* are the second highest overall, with *CSB2* and *AdaC2* having low scores. Of these two, *CSB2* has a slightly higher F-measure overall.

5   Discussion and Conclusion

Calibration, the act of adjusting the scores of classifiers so that they correspond to reliable probability estimates, is an often overlooked aspect of classification. We can use it to improve classification performance, especially in asymmetric situations. In the case of boosting, the results we obtained clearly show that *calibrated AdaBoost* can be used as a viable alternative to asymmetric versions of AdaBoost. In our experiments, we found that *calibrated AdaBoost* outperforms the asymmetric AdaBoost variants on datasets with large numbers of features while it performs comparably on datasets with few features. Furthermore, the theoretical properties of the original AdaBoost are preserved. Finally, we can also easily adjust our predictions without the need to retrain the model. This is also true for AdaBoost variants that modify only the prediction rule (*AdaMEC*), but not for those that modify the training phase (*CSB2, AdaC2*).

This study was limited to comparing *AdaMEC, CSB2* and *AdaC2* to *calibrated AdaBoost*, by virtue of being the most successful representatives of their families. As for the other variants, they are all methods that modify the training algorithm. *CSB0* and *CSB1* [17] do not use confidence rated predictions and based on the results of comparative studies [9,10,15], the two variants are typically dominated by *CSB2*. *Asymmetric-Adaboost* [18] was excluded from said studies as being similar to *CSB2*. AdaCost [2,17] is also outperformed by *AdaC2* and *CSB2* and so is *AdaC3* [16]. *CS-AdaBoost* [9,10], despite being the only method other than *AdaC2* with a solid theoretical basis, has been characterized as 'time-consuming and imprecise' [19], as it lacks a closed form solution for α_t and the optimization of its parameters is therefore computationally intensive.

To our knowledge, the only previous attempt at directly comparing asymmetric AdaBoost variants to *calibrated AdaBoost* was by Masnadi-Shirazi and Vasconselos [10]. The comparison was performed on imbalanced data, it included *AdaC2* and *CSB2* and the performance of *calibrated AdaBoost* was found to be slightly inferior to theirs. However the authors were solving a quite different problem from the one we do in the present paper. They *fixed the desired precision* of their ensembles and based on that they chose the appropriate cost setup for each method (i.e. the cost ratio c *differed* from one method to the other) so as to minimize the total number of errors on the test set.

On the other hand, we solve a cost-sensitive problem. We therefore use a *fixed* cost ratio c, taken directly from the cost matrix of the problem and our goal is to minimize the average cost of misclassifications. Our findings and those of Masnadi-Shirazi and Vasconselos are complementary, not contradictory. We observed that *AdaC2* and *CSB2* favor more aggressively the positive class compared to *calibrated AdaBoost*. Masnadi-Shirazi and Vasconselos, by fixing the precision to a high value, allow the ensemble to commit only a small number of false positives. *AdaC2* and *CSB2* are thus forced to select c values that limit their aggressiveness. Under this light, asymmetric AdaBoost variants can outperform *calibrated AdaBoost* on imbalanced data, if costs are allowed to vary, but with fixed costs, *calibrated AdaBoost* produces lower average costs.

Acknowledgments. This work was supported by EPSRC grant [EP/I028099/1]. We also thank Peter Flach for suggesting the idea that inspired this paper.

References

1. Elkan, C.: The foundations of cost-sensitive learning. In: IJCAI (2001)
2. Fan, W., Stolfo, S.J., Zhang, J., Chan, P.K.: Adacost: misclassification cost-sensitive boosting. In: ICML, pp. 97–105 (1999)
3. Flach, P.A.: The geometry of roc space: understanding machine learning metrics through roc isometrics. In: AAAI, pp. 194–201 (2003)
4. Freund, Y., Schapire, R.E.: A decision-theoretic generalization of on-line learning and an application to boosting. J. Comp. Syst. Sci. **55**(1), 119–139 (1997)
5. Friedman, J., Hastie, T., Tibshirani, R.: Additive logistic regression: a statistical view of boosting. Ann. Stat. **28**, 337–407 (2000)
6. Joshi, M.V., Kumar, V., Agarwal, R.C.: Evaluating boosting algorithms to classify rare classes: comparison and improvements (2001)
7. Landesa-Vázquez, I., Alba-Castro, J.L.: Shedding light on the asymmetric learning capability of adaboost. Pat. Rec. Lett. **33**(3), 247–255 (2012)
8. Maloof, M.A.: Learning when data sets are imbalanced and when costs are unequal and unknown. In: ICML (2003)
9. Masnadi-Shirazi, H., Vasconcelos, N.: Asymmetric boosting. In: ICML, pp. 609–619 (2007)
10. Masnadi-Shirazi, H., Vasconcelos, N.: Cost-sensitive boosting. IEEE Trans. Pat. Anal. Mach. Intell. **33**(2), 294–309 (2011)
11. Niculescu-Mizil, A., Caruana, R.: Obtaining calibrated probabilities from boosting. In: UAI (2005)
12. Platt, J.C.: Probabilistic outputs for support vector machines and comparisons to regularized likelihood methods. In: Advances in Large Margin Classifiers, pp. 61–74. MIT Press (1999)
13. Robertson, T., Wright, F., Dykstra, R.: Order Restricted Statistical Inference. Probability and Statistics. Wiley, New York (1988)
14. Schapire, R.E., Singer, Y.: Improved boosting algorithms using confidence-rated predictions. Mach. Learn. **37**(3), 297–336 (1999)
15. Sun, Y., Kamel, M.S., Wong, A.K.C., Wang, Y.: Cost-sensitive boosting for classification of imbalanced data. Pat. Recogn. **40**(12), 3358–3378 (2007)

16. Sun, Y., Wong, A.K.C., Wang, Y.: Parameter inference of cost-sensitive boosting algorithms. In: Perner, P., Imiya, A. (eds.) MLDM 2005. LNCS (LNAI), vol. 3587, pp. 21–30. Springer, Heidelberg (2005)
17. Ting, K.M.: A comparative study of cost-sensitive boosting algorithms. In: ICML, pp. 983–990 (2000)
18. Viola, P., Jones, M.: Fast and robust classification using asymmetric adaboost and a detector cascade. In: NIPS (2002)
19. Wang, Z., Fang, C., Ding, X.: Asymmetric real adaboost. In: ICPR (2008)
20. Zadrozny, B., Elkan, C.: Obtaining calibrated probability estimates from decision trees and naive bayesian classifiers. In: ICML, pp. 609–616 (2001)
21. Zadrozny, B., Elkan, C.: Transforming classifier scores into accurate multiclass probability estimates (2002)

Building Classifier Ensembles Using Greedy Graph Edit Distance

Kaspar Riesen[1]([envelope]), Miquel Ferrer[1], and Andreas Fischer[2]

[1] Institute for Information Systems, University of Applied Sciences and Arts
Northwestern Switzerland, Riggenbachstrasse 16, 4600 Olten, Switzerland
{kaspar.riesen,miquel.ferrer}@fhnw.ch
[2] Department of Informatics, University of Fribourg, Boulevard de Perolles 90,
1700 Fribourg, Switzerland
andreas.fischer@unifr.ch

Abstract. Classifier ensembles aim at more accurate classifications than single classifiers. In the present paper we introduce a general approach to building structural classifier ensembles, i.e. classifiers that make use of graphs as representation formalism. The proposed methodology is based on a recent graph edit distance approximation. The major observation that motivates the use of this particular approximation is that the resulting distances crucially depend on the order of the nodes of the underlying graphs. Our novel methodology randomly permutes the node order N times such that the procedure leads to N different distance approximations. Next, a distance based classifier is trained for each approximation and the results of the individual classifiers are combined in an appropriate way. In several experimental evaluations we make investigations on the classification accuracy of the resulting classifier ensemble and compare it with two single classifier systems.

1 Introduction

Classification is a common task in the area of pattern recognition and related fields. In order to compensate errors of a single classifier, the use of *classifier ensembles*, also referred to as *multiple classifier systems*, turns out to be a rewarding avenue to be pursued in many applications [1]. In particular, if the sets of misclassified patterns by the different classifiers of an ensemble do not heavily overlap, the classification accuracy of a classifier ensemble is nearly always beneficial in terms of the resulting classification accuracy.

Due to the mathematical wealth of operations available in a vector space, a huge amount of algorithms for classification of patterns formally represented by feature vectors have been developed in recent years [2,3]. Moreover, a large number of methods for the creation and combination of vector based classifiers have been developed (such as *bagging* [4], *boosting* [5,6], or feature subset selection [7]).

The use of feature vectors for pattern representation, however, implicates two severe limitations. First, vectors always represent a predefined set of features,

© Springer International Publishing Switzerland 2015
F. Schwenker et al. (Eds.): MCS 2015, LNCS 9132, pp. 125–134, 2015.
DOI: 10.1007/978-3-319-20248-8_11

and thus, all vectors in a particular application have to preserve the same length regardless of the size or complexity of the corresponding pattern. Second, there is no direct possibility to describe relationships among different parts of a pattern. Both constraints can be overcome by graph based representations. That is, graphs are able to explicitly describe binary relationships, and the number of nodes and edges of a graph can be adapted to each individual pattern. Due to the high representational power and the flexibility of graphs a growing interest in this representation formalism can be observed [8,9].

Graphs have been also used in the context of multiple classifier systems. A pioneering paper is [10] where it is shown that using vectors and graphs in an ensemble significantly improves the accuracy of a fingerprint recognition system. In [11] several graph representations of the same pattern are derived and merged into a single representation format. In [12], random node selection on graphs is used in order to derive classifier ensembles. Finally, in [13] graph embedding in real vector spaces by means of randomized prototype selection is used for building a general multiple classifier system for graphs.

The availability of a distance measure is a basic requirement for many (multiple) classifier systems. A large number of procedures for the computation of graph dissimilarity, commonly referred to as *graph matching*, have been proposed (see [14,15] for exhaustive surveys). *Graph edit distance* [16,17], introduced about 30 years ago, is still one of the most flexible and versatile graph matching models available. In particular, graph edit distance is able to cope with directed and undirected, as well as with labeled and unlabeled graphs. Additionally, if there are labels on nodes, edges, or both, no constraints on the respective label alphabets have to be considered.

Yet, a major drawback of graph edit distance is its computational complexity which is exponential in the number of nodes. In a recent publication [18] the authors of the present paper introduced an algorithmic framework for the approximation of graph edit distance in quadratic time. The basic idea of this approach is to reduce the difficult problem of graph edit distance computation to an assignment problem of local graph structures. A major characteristic of this novel framework is that the resulting approximation depends on the order in which the nodes of both graphs are processed. In other words, the resulting distance can be varied by permuting the node order of the graphs. The novel methodology of the present paper exploits this characteristic. In particular we aim at using the novel approximation framework in conjunction with random node permutations in order to produce diverse distance approximations and eventually combine these distances in a classifier ensemble.

The remainder of this paper is organized as follows. Next, in Sect. 2 the concept of graph edit distance as well as the recent framework for graph edit distance approximation [18] are summarized. In Sect. 3 the building of the classifier ensemble is described. An experimental evaluation on three real world data sets is carried out in Sects. 4 and 5 we draw conclusions and discuss several options for future work.

2 Graph Edit Distance

2.1 Basic Definitions

A graph g is a four-tuple $g = (V, E, \mu, \nu)$, where V is the finite set of nodes, $E \subseteq V \times V$ is the set of edges, $\mu : V \to L_V$ is the node labeling function, and $\nu : E \to L_E$ is the edge labeling function. The labels for both nodes and edges can be given by the set of integers $L = \{1, 2, 3, \ldots\}$, the vector space $L = \mathbb{R}^n$, a set of symbolic labels $L = \{\alpha, \beta, \gamma, \ldots\}$, or a combination of various label alphabets from different domains. Unlabeled graphs are obtained as a special case by assigning the same (empty) label $\varnothing$ to all nodes and edges, i.e. $L_V = L_E = \{\varnothing\}$.

Given two graphs, the source graph $g_1 = (V_1, E_1, \mu_1, \nu_1)$ and the target graph $g_2 = (V_2, E_2, \mu_2, \nu_2)$, the basic idea of graph edit distance [16,17] is to transform g_1 into g_2 using some edit operations. A standard set of edit operations is given by *insertions*, *deletions*, and *substitutions* of both nodes and edges. We denote the substitution of two nodes $u \in V_1$ and $v \in V_2$ by $(u \to v)$, the deletion of node $u \in V_1$ by $(u \to \varepsilon)$, and the insertion of node $v \in V_2$ by $(\varepsilon \to v)$, where ε refers to the empty node. For edge edit operations we use a similar notation.

A sequence $(e_1, \ldots, e_k)$ of k edit operations e_i that transform g_1 completely into g_2 is called *edit path* $\lambda(g_1, g_2)$ between g_1 and g_2. Let $\Upsilon(g_1, g_2)$ denote the set of all admissible edit paths between two graphs g_1 and g_2. To find the most suitable edit path out of $\Upsilon(g_1, g_2)$, one commonly introduces a cost $c(e)$ for every edit operation e, measuring the strength of the corresponding operation. The idea of such a cost is to define whether or not an edit operation e represents a strong modification of the graph.

Clearly, between two similar graphs, there should exist an inexpensive edit path, representing low cost operations, while for dissimilar graphs an edit path with high cost is needed. Consequently, the *edit distance* $d_{\lambda_{\min}}(g_1, g_2)$, or $d_{\lambda_{\min}}$ for short, of two graphs g_1 and g_2 is defined as

$$d_{\lambda_{\min}}(g_1, g_2) = \min_{\lambda \in \Upsilon(g_1, g_2)} \sum_{e_i \in \lambda} c(e_i). \tag{1}$$

2.2 Graph Edit Distance as Assignment Problem

Considering m nodes in g_1 and n nodes in g_2, $\Upsilon(g_1, g_2)$ contains $O(m^n)$ edit paths to be explored, and thus the computational complexity of exact graph edit distance is exponential. This means that for large graphs the computation of edit distance is intractable. The graph edit distance approximation framework originally introduced in [19] and extended in [18] reduces the difficult problem of graph edit distance computation to an instance of a *Linear Sum Assignment Problem (LSAP)* for which a large number of efficient algorithms exist [20]. The LSAP is defined as follows.

Definition 1. *Given two disjoint sets $S = \{s_1, \ldots, s_n\}$ and $Q = \{q_1, \ldots, q_n\}$ and an $n \times n$ cost matrix $\mathbf{C} = (c_{ij})$, where c_{ij} measures the cost of assigning the* *i-th element of the first set to the j-th element of the second set, the Linear Sum Assignment Problem (LSAP) consists in finding the minimum cost permutation*

$$(\varphi_1, \ldots, \varphi_n) = \underset{(\varphi_1, \ldots, \varphi_n) \in \mathcal{S}_n}{\arg \min} \sum_{i=1}^{n} c_{i\varphi_i},$$

where $\mathcal{S}_n$ refers to the set of all n! possible permutations of n integers, and permutation $(\varphi_1, \ldots, \varphi_n)$ refers to the assignment where the first entity $s_1 \in S$ is mapped to entity $q_{\varphi_1} \in Q$, the second entity $s_2 \in S$ is assigned to entity $q_{\varphi_2} \in Q$, and so on.

By reformulating the graph edit distance problem to an instance of an LSAP, three major issues have to be resolved. First, LSAPs are generally stated on independent sets with equal cardinality. Yet, in our case the elements to be assigned to each other are given by the sets of nodes (and edges) with unequal cardinality in general. Second, solutions to LSAPs refer to assignments of elements in which every element of the first set is assigned to exactly one element of the second set and vice versa (i.e. a solution to an LSAP corresponds to a bijection). Yet, graph edit distance is a more general assignment problem as it explicitly allows both deletions and insertions to occur on the basic entities (rather than only substitutions). Third, graphs do not only consist of independent sets of entities (i.e. nodes) but also of structural relationships between these entities (i.e. edges that connect pairs of nodes). LSAPs are not able to consider these relationships in a global and consistent way. The first two issues are perfectly – and the third issue partially – resolvable by means of the following definition of a square cost matrix whereon the LSAP is eventually solved.

Definition 2. *Based on the node sets $V_1 = \{u_1, \ldots, u_n\}$ and $V_2 = \{v_1, \ldots, v_m\}$ of g_1 and g_2, respectively, a cost matrix $\mathbf{C}$ is established as follows.*

$$\mathbf{C} = \begin{bmatrix} c_{11} & c_{12} & \cdots & c_{1m} & c_{1\varepsilon} & \infty & \cdots & \infty \\ c_{21} & c_{22} & \cdots & c_{2m} & \infty & c_{2\varepsilon} & \ddots & \vdots \\ \vdots & \vdots & \ddots & \vdots & \vdots & \ddots & \ddots & \infty \\ c_{n1} & c_{n2} & \cdots & c_{nm} & \infty & \cdots & \infty & c_{n\varepsilon} \\ c_{\varepsilon 1} & \infty & \cdots & \infty & 0 & 0 & \cdots & 0 \\ \infty & c_{\varepsilon 2} & \ddots & \vdots & 0 & 0 & \ddots & \vdots \\ \vdots & \ddots & \ddots & \infty & \vdots & \ddots & \ddots & 0 \\ \infty & \cdots & \infty & c_{\varepsilon m} & 0 & \cdots & 0 & 0 \end{bmatrix} \qquad (2)$$

Entry c_{ij} thereby denotes the cost of a node substitution $(u_i \to v_j)$, $c_{i\varepsilon}$ denotes the cost of a node deletion $(u_i \to \varepsilon)$, and $c_{\varepsilon j}$ denotes the cost of a node insertion $(\varepsilon \to v_j)$.

Note that matrix $\mathbf{C} = (c_{ij})$ is by definition quadratic. Hence, the first issue (sets of unequal size) is instantly eliminated. Obviously, the left upper corner of the cost matrix $\mathbf{C} = (c_{ij})$ represents the costs of all possible node substitutions, the diagonal of the right upper corner the costs of all possible node deletions, and the diagonal of the bottom left corner the costs of all possible node insertions. Note that every node can be deleted or inserted at most once. Therefore any non-diagonal element of the right-upper and left-lower part is set to ∞. The bottom right corner of the cost matrix is set to zero since substitutions of the form $(\varepsilon \to \varepsilon)$ should not cause any cost.

Given the cost matrix $\mathbf{C} = (c_{ij})$, the assignment problem consists in finding a permutation $(\varphi_1, \ldots, \varphi_{n+m})$ of the integers $(1, 2, \ldots, (n + m))$ that minimizes the overall assignment cost $\sum_{i=1}^{(n+m)} c_{i\varphi_i}$. This permutation corresponds to the assignment

$$\psi = ((u_1 \to v_{\varphi_1}), (u_2 \to v_{\varphi_2}), \ldots, (u_{m+n} \to v_{\varphi_{m+n}}))$$

of the nodes of g_1 to the nodes of g_2. Note that assignment ψ includes node assignments of the form $(u_i \to v_j)$, $(u_i \to \varepsilon)$, $(\varepsilon \to v_j)$, and $(\varepsilon \to \varepsilon)$ (the latter can be dismissed, of course). Hence, the definition of the cost matrix in Eq. 2 also resolves the second issue stated above and allows insertions and/or deletions to occur in an optimal assignment.

The third issue is about considering the edge structure of both graphs, and it can only partially be resolved. So far the the cost matrix $\mathbf{C} = (c_{ij})$ considers the nodes of both graphs only, and thus assignment ψ does not take any structural constraints into account. In order to integrate knowledge about the graph structure, to each entry $c_{ij} \in \mathbf{C}$ the minimum sum of edge edit operation costs, implied by the corresponding node operation, is added. That is, we encode the minimum cost arising from the local edge structure in the individual entries $c_{ij} \in \mathbf{C}$, which enables the consideration of information about the local, yet not global, edge structure of a graph.

Given the node assignment ψ a distance value approximating the exact graph edit distance $d_{\lambda_{\min}}$ can be directly inferred. Note that edit operations on edges are always defined by the edit operations on their adjacent nodes. That is, whether an edge (u, v) is substituted, deleted, or inserted, depends on the edit operations actually performed on both adjacent nodes u and v. Hence, the edge operations can be completely inferred from ψ (which contains a complete and consistent set of node edit operations). Hence, we finally get an admissible edit path from $\Upsilon(g_1, g_2)$ and a corresponding approximate edit distance $d_\psi(g_1, g_2)$, or d_ψ for short.

Note that the edit path corresponding to $d_\psi(g_1, g_2)$ considers the edge structure of g_1 and g_2 in a global and consistent way while the optimal node assignment ψ is able to consider the structural information in an isolated way only (single nodes and their adjacent edges). Therefore, the distance d_ψ found by this specific framework is – in the best case – equal to, or – in general – larger than the exact graph edit distance $d_{\lambda_{\min}}$.

2.3   Greedy Graph Edit Distance (Greedy-GED)

The computation of a permutation $(\varphi_1, \ldots, \varphi_{n+m})$ and the corresponding assignment ψ is the core process of the complete approximation procedure. This task corresponds to an instance of an LSAP and thus a large number of optimal algorithms exist (see [20] for an exhaustive survey). The time complexity of the best performing optimal algorithms for LSAPs is cubic in the size of the problem, which corresponds to $O((n + m)^3)$ in our case. In the original framework [19] the LSAP stated on $\mathbf{C}$ is optimally solved by means of Munkres' algorithm [21]. In a recent extension [18] a suboptimal algorithm is used to find an assignment ψ on the basis of $\mathbf{C}$.

The idea of this suboptimal assignment is formalized in Algorithm 1. This algorithm iterates through cost matrix $\mathbf{C}$ from top to bottom through all rows and assigns every element to the minimum unused element in a greedy manner. More formally, for each row i in the cost matrix $\mathbf{C} = (c_{ij})$ the minimum cost entry $\varphi_i = \arg\min_{\forall j} c_{ij}$ is determined and the corresponding node edit operation $(u_i \to v_{\varphi_i})$ is added to ψ. By removing column φ_i in $\mathbf{C}$ it is ensured that every column of the cost matrix is considered exactly once (i.e. $\forall j$ refers to available columns in $\mathbf{C}$). In the first row we have to consider $(n + m)$ elements in order to find the minimum. In the second row we have to consider $(n + m - 1)$ elements, a.s.o. In the last row only one element remains. Hence, the complexity of this approximate assignment algorithm is $O((n + m)^2)$.

Algorithm 1. Greedy-Assignment($\mathbf{C} = (c_{ij})$)

1: $\psi = \{\}$
2: **for** $i = 1, \dots, (m + n)$ **do**
3:     $\varphi_i = \arg\min_{\forall j} c_{ij}$
4:         Remove column φ_i from $\mathbf{C}$
5:     $\psi = \psi \cup \{(u_i \to v_{\varphi_i})\}$
6: **end for**

7: **return** ψ

In contrast with an optimal permutation $(\varphi_1, \varphi_2, \dots, \varphi_{(n+m)})$ returned by the original framework [19], the permutation $(\varphi'_1, \varphi'_2, \dots, \varphi'_{(n+m)})$ returned by Algorithm 1 is suboptimal. That is, the sum of assignments costs of our greedy approach is greater than, or equal to, the minimal assignment cost provided by optimal LSAP solving algorithms:

$$\sum_{i=1}^{(n+m)} c_{i\varphi'_i} \geq \sum_{i=1}^{(n+m)} c_{i\varphi_i}$$

Yet, note that for the corresponding distance values d_ψ and $d_{\psi'}$ no globally valid order relation exists (ψ and ψ' correspond to the optimal and greedy permutation, respectively). That is, the approximate graph edit distance $d_{\psi'}$ derived from ψ' can be greater than, equal to, or smaller than d_ψ.

For the remainder of this paper we denote this algorithmic procedure with *Greedy-GED*.

3   Classifier Ensemble Based on Greedy-GED

Note that the nodes of a graph are generally unordered and thus, the first n rows (and m columns) in $\mathbf{C}$ are arbitrarily ordered. Note, however, that the assignment method described in Algorithm 1 operates in a greedy manner, i.e. it is not able to undo a certain node assignment once it has been added to ψ. Hence, this method and in particular the resulting distance approximation d_ψ crucially depends on the order in which the nodes are processed.

In [22] this clear drawback is partially resolved by some refinements in the greedy assignment process. The present paper, however, exploits the drawback of the original procedure as an advantage in order to build a classifier ensemble. In particular, before the greedy assignment of Algorithm 1 is actually carried out, we randomly permute the order of the rows in $\mathbf{C}$. Clearly, this procedure can be used to produce N potentially different assignments $\psi^1, \psi^2, \ldots, \psi^N$ from the same cost matrix $\mathbf{C}$. These assignments in turn might lead to N different graph edit distance approximations $d_{\psi^1}, d_{\psi^2}, \ldots, d_{\psi^N}$.

The N distances obtained by this randomized procedure are used in conjunction with a nearest neighbor classifier. Note that there are other approaches to graph classification available that make use of graph edit distance in some form, including vector space embedding classifiers [23] and graph kernels [24]. Yet, the nearest neighbor paradigm is particularly interesting for the present evaluation because it directly uses the distances without any additional classifier training.

For the final classification based on the N decisions a *plurality voting* is carried out. That is, the class label of the nearest neighbor output by the classifier that makes use of the i-th randomized graph edit distance approximation is regarded as one vote for this particular class ($i = 1, \ldots, N$). The class that receives the plurality of the votes is choosen by the combiner. Of course, The proposed procedure is basically applicable with more restrictive voting methods (e.g. *majority voting* [1]) as well as more elaborated combining methods (see Sect. 5 for a more detailed discussion on possible extensions).

4   Experimental Evaluation

For our empirical investigations we use two data sets from the IAM graph database repository [25] and one data set from GREYC's data set repository[1]. All data sets involve graphs that represent molecular compounds and all sets consists of two classes. In Table 1 the main characteristics of the three data sets are summarized[2]. For more details on the graph extraction methods and the graph characteristics we refer to [25].

The purpose of the experiments described in this section is to compare the classification accuracy of the ensemble obtained by the proposed method with

---

[1] https://brunl01.users.greyc.fr/CHEMISTRY/index.html.

[2] For the MAO data only a small training set is available and thus we conduct a *leave-one out* experiment on this data set.

Table 1. The size of the training and test set $|tr|$ and $|te|$, respectively, and the mean and max number of nodes and edges $|V|$ and $|E|$, respectively.

Data	$|tr|$	$|te|$	$\varnothing|V|$	$\varnothing|E|$	max $|V|$	max $|E|$
AIDS	250	1500	15.7	16.2	95	103
MUTA	1000	1000	30.3	30.8	417	112
MAO	70	–	18.4	19.6	27	29

Table 2. Recognition accuracy in percentage of the proposed ensemble and the two reference systems.

Data	Reference system		Ensemble
	$\mu \pm \sigma$	max	Voting
AIDS	98.85 ± 0.11	99.00	99.00
MUTA	66.14 ± 0.96	67.70	68.70
MAO	67.64 ± 4.53	73.53	76.47

the mean and maximum accuracy of all individual ensemble members (referred to as μ and max, respectively). We set $N = 9$ for our evaluation.

In Table 2 the recognition accuracies of our novel ensemble and the reference systems are shown (note that we additionally indicate the standard deviation σ of the ensemble accuracies). We observe that the ensemble outperforms the mean accuracy of the individual members on all data sets. Moreover, in two out of three cases the ensemble outperforms the second reference system (the best performing individual classifier per ensemble). That is, we can conclude that our novel approach for building graph based classifier ensembles is clearly beneficial.

The standard deviation σ of the accuracy on the MUTA and MAO data set indicate quite a large variation of the individual classifications. Moreover, on the MAO data set at least one classifier from the ensemble returns a correct classification in more than 98 % of all cases. The gap between this result and the accuracy of our ensemble (76.47 %) confirms the great variety of classification results among all ensemble members. Similar results are observable on the MUTA data sets, while the results from the set of misclassified patterns do not heavily overlap on the AIDS data set (at least one ensemble member correctly classifies the graphs in 99.20 % and 93.10 % of all cases on the AIDS and MUTA data sets, respectively).

5   Conclusions and Future Work

The present paper introduces a novel ensemble generation method applicable to structural pattern representation (in particular to graphs). The basic idea is to use a greedy graph edit distance approximation which returns different distance approximations depending on the node order of the underlying graphs. Hence, by randomly permuting the nodes of the first graph one can instantly derive different distance approximations for the same graph pair. Any distance based classifier can eventually be used to build a multiple classifier system based on these diverse distances. In the present paper an ensemble of nearest neighbor classifiers is generated and tested on three real world data sets.

The experimental evaluation is only a very first step towards a better understanding of the advantages and limitations of our novel system. However, the results are convincing and clearly indicate that this line of research is worth to be pursued in future work. Among others we identify three possible extensions of our basic system. First, we plan to substantially increase the number of individual classifiers to some hundreds and then apply search strategies (such as a *floating search* or similar) in order to find an optimal subset of classifiers to be used as ensemble (known as *overproduce-and-select*). Second, we aim at investigating whether elaborated variants of the greedy assignment might help in building better individual classifiers. The third extension consists in using other combination strategies to obtain the final classification results (such as *Borda count* or *Bayes' combination* using a plausibility function derived from the distance approximations).

Acknowledgements. This work has been supported by the Swiss National Science Foundation (SNSF) projects Nr. 200021_153249 and P300P2_1512 as well as the Hasler Foundation Switzerland.

References

1. Kuncheva, L.: Combining Pattern Classifiers: Methods and Algorithms. Wiley, New Jersey (2004)
2. Bishop, C.: Pattern Recognition and Machine Learning. Springer, New York (2008)
3. Shawe-Taylor, J., Cristianini, N.: Kernel Methods for Pattern Analysis. Cambridge University Press, Cambridge (2004)
4. Breiman, L.: Bagging predictors. Mach. Learn. **24**, 123–140 (1996)
5. Freund, Y., Shapire, R.: A decision theoretic generalization of online learning and application to boosting. J. Comput. Syst. Sci. **55**, 119–139 (1997)
6. Shapire, R., Freund, Y., Bartlett, P., Lee, W.: Boosting the margin: a new explanation for the effectiveness of voting methods. Ann. Stat. **26**(5), 1651–1686 (1998)
7. Ho, T.: The random subspace method for constructing decision forests. IEEE Trans. Pattern Anal. Mach. Intell. **20**(8), 832–844 (1998)
8. Cook, D., Holder, L. (eds.): Mining Graph Data. Wiley-Interscience, New York (2007)
9. Gärtner, T., Horvath, T., Wrobel, S.: Graph kernels. Encycl. Mach. Learn. **2010**, 467–469 (2010)
10. Marcialis, G., Roli, F., Serrau, A.: Fusion of statistical and structural fingerprint classifiers. In: Kittler, J., Nixon, M.S. (eds.) AVBPA 2003. LNCS, vol. 2688, pp. 310–317. Springer, Heidelberg (2003)
11. Neuhaus, M., Bunke, H.: Graph-based multiple classifier systems a data level fusion approach. In: Roli, F., Vitulano, S. (eds.) ICIAP 2005. LNCS, vol. 3617, pp. 479–486. Springer, Heidelberg (2005)

12. Schenker, A., Bunke, H., Last, M., Kandel, A.: Building graph-based classifier ensembles by random node selection. In: Roli, F., Kittler, J., Windeatt, T. (eds.) MCS 2004. LNCS, vol. 3077, pp. 214–222. Springer, Heidelberg (2004)

13. Riesen, K., Bunke, H.: Classifier ensembles for vector space embedding of graphs. In: Haindl, M., Kittler, J., Roli, F. (eds.) MCS 2007. LNCS, vol. 4472, pp. 220–230. Springer, Heidelberg (2007)

14. Conte, D., Foggia, P., Sansone, C., Vento, M.: Thirty years of graph matching in pattern recognition. Int. J. Pattern Recogn. Art Intelligence 18(3), 265–298 (2004)

15. Foggia, P., Percannella, G.: Graph matching and learning in pattern recognition in the last 10 years. Int. J. Pattern Recogn. Art Intell. 28(1), 40 (2014)

16. Sanfeliu, A., Fu, K.: A distance measure between attributed relational graphs for pattern recognition. IEEE Trans. Syst. Man Cybern. (Part B) 13(3), 353–363 (1983)

17. Bunke, H., Allermann, G.: Inexact graph matching for structural pattern recognition. Pattern Recogn. Lett. 1, 245–253 (1983)

18. Riesen, K., Ferrer, M., Dornberger, R., Bunke, H.: Greedy graph edit distance (2015). Submitted to MLDM

19. Riesen, K., Bunke, H.: Approximate graph edit distance computation by means of bipartite graph matching. Image Vis. Comput. 27(4), 950–959 (2009)

20. Burkard, R., Dell'Amico, M., Martello, S.: Assignment Problems. Society for Industrial and Applied Mathematics, Philadelphia (2009)

21. Munkres, J.: Algorithms for the assignment and transportation problems. J. Soc. Indus. Appl. Math. 5(1), 32–38 (1957)

22. Riesen, K., Ferrer, M., Fischer, A., Bunke, H.: Approximation of graph edit distance in quadratic time (2015). Submitted to GbR

23. Riesen, K., Bunke, H.: Graph classification based on vector space embedding. Int. J. Pattern Recogn. Artif. Intell. 23(6), 1053–1081 (2008)

24. Neuhaus, M., Bunke, H.: Bridging the Gap Between Graph Edit Distance and Kernel Machines. World Scientific, Switzerland (2007)

25. Riesen, K., Bunke, H.: IAM graph database repository for graph based pattern recognition and machine learning. In: da Vitoria, L., et al. (eds.) Structural, Syntactic, and Statistical Pattern Recognition. LNCS, vol. 5342, pp. 287–297. Springer, Heidelberg (2008)

Measuring the Stability of Feature Selection with Applications to Ensemble Methods

Sarah Nogueira$^{(\boxtimes)}$ and Gavin Brown

School of Computer Science, University of Manchester, Manchester M13 9PL, UK
{sarah.nogueira,gavin.brown}@manchester.ac.uk

Abstract. Ensemble methods are often used to decide on a good selection of features for later processing by a classifier. Examples of this are in the determination of Random Forest *variable importance* proposed by Breiman, and in the concept of *feature selection ensembles*, where the outputs of multiple feature selectors are combined to yield more robust results. All of these methods rely critically on the concept of feature selection *stability* - similar but distinct to the concept of *diversity* in classifier ensembles. We conduct a systematic study of the literature, identifying desirable/undesirable properties, and identify a weakness in existing measures. A simple correction is proposed, and empirical studies are conducted to illustrate its utility.

Keywords: Stability · Feature selection · Ensembles

1 Introduction

The stability of feature selection can be seen as its sensitivity to *small changes* in the input dataset. In many applications, stability of feature selection is crucial as the user might need to identify an interpretable feature subset, e.g. when identifying genes responsible for a disease [1]. Stable feature selection frameworks provide more reliable feature subsets and gain in interpretability. As the output of a feature selection algorithm (FSA) can either be a set of features, a ranking on the features or a score on the features, there exist stability measures that apply to each one of these cases. In this paper, we focus on the first case where an FSA returns a feature set.

Why is measuring stability of feature selection an issue? In regression or classification predictors, the sensitivity to changes in data is quantified exactly in a *bias-variance* decomposition of the error measure (though in the classification case this is not entirely straightforward). There is no such decomposition that applies to feature selection. First of all, the *true* relevant set of features is unknown (and strongly depends on the classifier that will be used afterwards) which does not allow us to define the concept of bias. In the case of regression predictors, such decomposition relies on the convexity of the squared-loss function. A stability measure will allow us to quantify the variability in the feature sets selected by an FSA for a given dataset.

© Springer International Publishing Switzerland 2015
F. Schwenker et al. (Eds.): MCS 2015, LNCS 9132, pp. 135–146, 2015.
DOI: 10.1007/978-3-319-20248-8_12

Why do we need a new measure? Kuncheva [8] demonstrated the importance of the property of *correction for chance* and derived a new measure satisfying this property. Nevertheless, the measure proposed can only be calculated for FSAs selecting a fixed amount of features on a given dataset. Even though several variants of Kuncheva's measure satisfying the property of correction for chance have been proposed to deal with feature sets of varying cardinality [9,12,14], we will show that they are flawed in the sense that they do not satisfy other critical properties (e.g. they do not always return their maximal value when the FSA always returns the same feature set or they are not bounded by constants). Hence, we derived a generalization of Kuncheva's stability measure that can be used with feature sets of varying cardinality while retaining a set of desirable properties. Examples that illustrate of the utility of these measures are feature selection techniques using hypothesis testing, random forests or LASSO. Indeed, when applying LASSO to different samples of the same data, there is no guarantee that the same coefficients will be equal to 0 and hence, that a constant number of features will be selected when different samples of the data are taken.

Applications to Ensemble-Based Feature Selection. Stability of ensemble-based feature selection has recently become area of interest [1,3,5,10]. In ensemble-based feature selection, we use a set of *diverse* feature selection methods to build a more *robust* one. A stability analysis could then be carried out to observe the diversity (corresponding to low stability) of the different feature selection methods within an ensemble as well as to observe the robustness (corresponding to high stability) of the feature selection made by the ensemble.

The remainder of the paper is structured as follows. Section 2 presents some of the properties of the existing measures. Section 3 focuses on the measures having the property of correction for chance for feature sets of different cardinalities and highlights their weaknesses on toy examples. Section 4 proposes a new measure having a set of identified properties and Sect. 5 illustrates its utility in the context of an ensemble-based feature selection procedure.

2   Stability Measures

2.1   Existing Measures

To observe the robustness of an FSA to changes in the data, the FSA is applied to K samples of the same dataset to obtain a sequence $\mathcal{A}$ of K feature sets. The more similar these K feature sets will be, the more the procedure will be said to be stable. To define stability, one common approach consists of defining a *similarity measure* sim between two feature sets s_1 and s_2 and then to define the stability as the average similarity over all pairs of feature sets in $\mathcal{A}$. In that case, the stability will be denoted by $\overline{sim}$ and we can express it as follows:

$$\overline{sim}(\mathcal{A}) = \frac{2}{K(K-1)} \sum_{i=1}^{K-1} \sum_{j=i+1}^{K} sim(s_i, s_j), \tag{1}$$

where s_i is the ith feature set in $\mathcal{A}$. Several similarity measures have been proposed in the literature. Some of the older works propose using the *Jaccard index* sim_J [6] (also referred as the *Tanimoto distance*) or the relative *Hamming distance* to define a similarity measure sim_H [4]. Let us assume that we have n features in total. The output of an FSA can then be seen as a binary string of length n with a 1 at the ith position if the ith feature has been selected and with a 0 otherwise. Let's assume that an FSA returns the following sequence $\mathcal{A}$ of $K = 3$ feature sets:

$$s_1: \ 1\,0\,0\,1\,0\,1 \qquad s_2: \ 1\,1\,0\,0\,0\,1 \qquad s_3: \ 1\,0\,1\,1\,1\,1 \qquad (2)$$

The similarity measure sim_H between two feature sets is defined as the number of bits they have in common divided by the length n of the string. Therefore, using Eq. 1, the resulting stability of these feature sets will be equal to:

$$\overline{sim_H}(\mathcal{A}) = \frac{2}{3(3-1)} \sum_{i=1}^{2} \sum_{j=i+1}^{3} sim_H(s_i, s_j) = \frac{2}{3(3-1)} \left(\frac{4}{6} + \frac{4}{6} + \frac{2}{6} \right) = \frac{5}{9}.$$
$$(3)$$

Nevertheless, both these measures are *subset-size-biased* [8] meaning that their values are biased by the number of features selected and hence cannot be used consistently to compare the stability of FSAs in different settings. Indeed, imagine that a procedure selects two identical feature sets of 8 features out of a total of 10 features and that another procedure selects two identical feature sets of 8 features out of a total of 100 features. Intuitively, the second procedure is more stable, as it is less likely to have selected the exact same 8 features by chance. For this reason, Kuncheva [8] proposed a similarity measure having the property of correction for chance. The similarity between two feature sets of size k can be seen as the number of features r they have in common (i.e. the size of their intersection). As we want this measure to reflect on the true ability of the procedure to select identical features, Kuncheva [8] proposes correcting r by the expected size of the intersection between two feature sets of k features drawn at random (denoted hereafter by $\mathbb{E}[r]$). The size of the intersection between two sets containing k objects each individually randomly drawn without replacement amongst a total of n objects follows a hypergeometric distribution, and therefore we have that $\mathbb{E}[r] = \frac{k^2}{n}$. In order to make this value comparable for different values of k and n, Kuncheva rescales $r - \mathbb{E}[r]$ in $[-1, 1]$ by dividing it by its maximal value $max(r - \mathbb{E}[r])$:

$$sim_K(s_1, s_2) = \frac{r - \mathbb{E}[r]}{max(r - \mathbb{E}[r])} = \frac{r - \mathbb{E}[r]}{max(r) - \mathbb{E}[r]} = \frac{r - \mathbb{E}[r]}{k - \mathbb{E}[r]} = \frac{r - \frac{k^2}{n}}{k - \frac{k^2}{n}}, \qquad (4)$$

where s_1 and s_2 are two feature sets of cardinality k and where $max(r)$ is the maximal possible value of r for a given k. The measure sim_K will hence reach its maximal value of 1 when $r = k$, i.e. when the two feature sets s_1 and s_2 are identical.

2.2   Properties

By leading a thorough study of the literature, we have identified the following set of desirable properties for a stability measure:

1. **Limits.** The measure should be bounded by values that do not depend on the number of features and the cardinality of the feature sets should reach its maximal value when the feature sets are identical.
2. **Monotonicity.** The measure should be an increasing function of the similarity of the feature sets.
3. **Correction for chance.** This property allows to compare stability of FSAs selecting a different amount of features. Positive values will be interpreted as being more stable than an FSA selecting features at random.
4. **Unconstrained on cardinality.** We would like a stability measure to be able to deal with feature sets of different cardinalities.
5. **Symmetry.** We would like the stability measure to be symmetrical, so that its value does not depend on the order on which the feature sets are taken.
6. **Redundancy awareness.** As features can be redundant, we would like a stability measure to reflect on the true amount of redundant information between the feature sets.

Properties 1 to 3 were the ones identified by Kuncheva [8] and Properties 4 to 6 are the ones that we have identified by looking at the measures proposed later on. Table 1 gives us the properties of the most commonly used existing stability measures for FSAs returning a feature set.

Table 1. Properties of stability measures for FSAs outputting a feature set.

	1	2	3	4	5	6
$\overline{sim}_J$ (Dunne et al. [4])	✓	✓		✓	✓	
$\overline{sim}_H$ (Kalousis et al. [6])	✓	✓		✓	✓	
$\overline{sim}_M$ (Yu et al. [13])	✓	✓		✓	✓	✓
$\overline{sim}_K$ (Kuncheva [8])	✓	✓	✓		✓	
$\overline{sim}_L$ (Lustgarten et al. [9])		✓	✓	✓	✓	
$\overline{sim}_W$ (Wald et al. [12])		✓	✓	✓	✓	
Average $nPOG$ (Zhang et al. [14])		✓	✓	✓		
Average $nPOGR$ (Zhang et al. [14])		✓	✓	✓		✓
CW_{rel} (Somol and Novovičová [11])	✓	✓		✓	✓	
γ_k (Krízek et al. [7])	✓	✓			✓	

The focus of this paper is on the stability measures having the important property of correction for chance. Even though the stability measure CW_{rel} (introduced Somol and Novovičová [11]) does not explicitly yield the property of correction for chance, using Theorem 1, we can point out that CW_{rel} asymptotically holds this property when a constant number of features is selected.

Theorem 1. *For a sequence $\mathcal{A}$ containing feature sets of constant cardinality, the stability measure $CW_{rel}(\mathcal{A})$ is asymptotically equivalent to $\overline{sim_K}(\mathcal{A})$ as the number of feature sets approaches infinity.*[1]

All the five stability measures having the property of correction for chance (cf Table 1) are either taken as the average pairwise similarities between the feature sets (as in $\overline{sim_K}$, $\overline{sim_L}$, $\overline{sim_W}$) or as the average similarity between disjoint feature sets pairs (for stability measures using $nPOG$ and $nPOGR$). The $nPOGR$ similarity measure is a generalization of the $nPOG$ measure that attempts to take into account linear feature redundancies (which is not in the scope of this paper). The similarity measures sim_L, sim_W and $nPOG$ are all variants of Kuncheva's similarity measure sim_K for feature sets of varying cardinalities.

3   Extensions of Kuncheva's Similarity Measure

3.1   Definitions

There are three similarity measures extending Kuncheva's similarity measure sim_K for feature sets s_1 and s_2 of different cardinalities (respectively k_1 and k_2). In this situation, the value of the expected size of the intersection for randomly drawn feature sets becomes $\mathbb{E}[r] = \frac{k_1 k_2}{n}$ [9]. The three measures are of the same general form as Kuncheva's measure sim_K, as they keep the numerator equal to $r - \mathbb{E}[r]$. In order to make these values comparable in different settings (i.e. for different values of k_1, k_2 and n), the value of $r - \mathbb{E}[r]$ needs to be rescaled. The three similarity measure extending sim_K are three variants of this and they only differ in the way the numerator $r - \mathbb{E}[r]$ is rescaled. Note that in all these expressions, the only variable is the size of the intersection r and that all other terms are constants only depending on k_1, k_2 and n. Lustgarten et al. [9] proposes dividing the value of the numerator by $r - \mathbb{E}[r]$ by its range (i.e. by its maximal value minus its minimal value for a given k_1, k_2 and n):

$$sim_L(s_1, s_2) = \frac{r - \mathbb{E}[r]}{\max(r - \mathbb{E}[r]) - \min(r - \mathbb{E}[r])}. \tag{5}$$

As $\mathbb{E}[r]$ is a constant only depending on k_1, k_2 and n, $r - \mathbb{E}[r]$ is a linear function of r and hence the above equation becomes:

$$sim_L(s_1, s_2) = \frac{r - \mathbb{E}[r]}{(\max(r) - \mathbb{E}[r]) - (\min(r) - \mathbb{E}[r])} = \frac{r - \mathbb{E}[r]}{\max(r) - \min(r)}, \tag{6}$$

where $max(r)$ and $min(r)$ are respectively the maximal and the minimal possible values of the size of the intersection r given k_1, k_2 and n. Intuitively, we can see that the minimal size of the intersection between two feature sets is not 0. Indeed, imagine we have a set containing $k_1 = 2$ features, another set containing $k_2 = 3$ features and that we have $n = 4$ features to select from in total. In this setting,

---

[1] Proofs of the theorems available at http://www.cs.man.ac.uk/~nogueirs.

these two sets cannot be disjoint. It can be shown that the minimal possible value of r is equal to $min(r) = max(0, k_1 + k_2 - n)$. Similarly the maximal value of r is reached when one set is a proper subset of the other. Therefore, the maximal value or r is equal to $max(r) = min(k_1, k_2)$. Lustgarten's measure sim_L can therefore be rewritten as follows:

$$sim_L(s_1, s_2) = \frac{r - \mathbb{E}[r]}{max(0, k_1 + k_2 - n) - min(k_1, k_2)}. \tag{7}$$

It can be shown that this rescaling procedure ensures a value of sim_L in the interval $[-1, 1]$. Nevertheless, we will see through a set of examples that this procedure does not satisfy all the desirable properties in this context.

In a similar way, Wald et al. [12] proposes rescaling the numerator by dividing it by its maximal value:

$$sim_W(s_1, s_2) = \frac{r - \mathbb{E}[r]}{max(r - \mathbb{E}[r])} = \frac{r - \mathbb{E}[r]}{max(r) - \mathbb{E}[r]} = \frac{r - \frac{k_1 k_2}{n}}{min(k_1, k_2) - \frac{k_1 k_2}{n}}. \tag{8}$$

By dividing the numerator by its maximal value, we are ensured that sim_W will always be less than or equal to 1. Nevertheless, as the numerator can take negative values, dividing it by the maximal value will not guarantee lower bounds that do not depend on the constants k_1, k_2 and n. In fact, it can be shown that for a given n, the minimum of sim_W is $1 - n$ (and is reached when $k_1 = n - 1$ and $k_2 = 1$ or vice versa). We will illustrate the importance of this with an example in the next Section. In the measure $nPOG$, Zhang et al. [14] divide the numerator either by $k_1 - \mathbb{E}[r]$ if s_1 is given as the first argument or by $k_2 - \mathbb{E}[r]$ otherwise; making the resulting similarity measure non-symmetrical (i.e. $nPOG(s_1, s_2) \neq nPOG(s_2, s_1)$):

$$nPOG(s_1, s_2) = \frac{r - \mathbb{E}[r]}{k_1 - \mathbb{E}[r]} = \frac{r - \frac{k_1 k_2}{n}}{k_1 - \frac{k_1 k_2}{n}}. \tag{9}$$

The non-symmetry of this measure can be problematic, as we will illustrate it in the next Section. Also, one can notice that when the set of smaller cardinality is given as the first argument, $nPOG$ is equal to the sim_W measure, hence inheriting its weaknesses.

3.2   Toy Examples Illustrating the Weaknesses of the Measures

To illustrate some of the missing properties of the similarity measures, we provide four toy examples.

Example 1: Accounting for Systematic Bias in Chosen Set Size. Imagine that there are 10 features to choose from. Procedure F_1 chooses 7 features *deterministically*, i.e. no matter what the variation in data is, the same 7 features are returned. Intuitively, the "stability" of this procedure is maximal. It has zero variation in its choice of feature set. Lustgarten's measure returns a value of $sim_L = 0.7$, whilst *all other measures* return 1. This is somewhat strange, and undesirable, as we have no way to know if F_1 is deterministic from the value of sim_L. Furthermore, imagine procedure F_2, which picks 4 features, again deterministically. Lustgarten's measure now returns 0.6, which makes procedures F_1 and F_2 not comparable in terms of stability. This example highlights the need for a similarity measure that returns its maximal value as long as the feature sets are identical, as stated in Property 1 of Sect. 2.

Example 2: Accounting for the Set Size Variations. Imagine our same set of 10 features as above. Half of the time, procedure F returns features 1 to 8, and the other half of the time it returns features 1 and 2, i.e. a proper subset. In this situation, Wald's stability measure returns a maximal 1, whilst clearly there is variation in the choice of the subset size. In fact, using Wald's measure, the similarity between two feature sets will always have its maximal value of 1 as long as one of the two sets is a proper subset of the other. For the other two similarity measures, this is not the case, even though the similarities do not decrease proportionally to the distance between the feature sets cardinalities.

Example 3: Invariance to Feature Set Permutations. This example is the same as the previous one where the order of the feature sets has been permuted. Because of the non-symmetry of $nPOG$, the stability value returned by this measure might not be the same as the one calculated in the previous example. This example showcases the need for a symmetrical similarity measure as stated in Property 5 of Sect. 2.

Example 4: Bounded by Constants. Having minimal values that increase linearly with n can lead to negative values of much larger amplitude than the maximum value which is equal to 1. If we have $n = 100$ features in total, the minimal value of $nPOG$ and of sim_W is equal to -99 while its maximal value is 1. When calculating the average of the similarities, such large negative values can strongly bias the resulting stability. We can illustrate this with a simple example. Imagine that a feature selection procedure selects 9 times features 1 to 8 in feature sets $s_1, s_2, ...s_9$ and that features 9 and 10 are selected in a set s_{10}. When averaging over all possible pairs of similarities, the stability value of $\overline{sim_W}$ is 0 which corresponds to the stability value of an FSA drawing 10 feature sets at random, even though 9 out of the 10 feature sets considered were identical. Another issue with minimal values depending on n is that the minimal values will be different for two different values of n. In practice, this does not allow us to compare the stability of an FSA on two different datasets for instance. This shows the need for a stability measure to be bounded by constants as stated in Property 1 of Sect. 2.

4   A New Similarity Measure

In the light of the previous observations, we propose a new similarity measure sim_N of the same general form that will rescale the numerator $r - \mathbb{E}[r]$ so that its value belongs to $[-1, 1]$. As the numerator $r - \mathbb{E}[r]$ can take both negative and positive values, one way to do so is to divide it by its maximal absolute value as follows:

$$sim_N(s_1, s_2) = \frac{r - \mathbb{E}[r]}{max\left(|r - \mathbb{E}[r]|\right)}. \tag{10}$$

Both Kuncheva's and Wald's similarity measures sim_K and sim_W rescale the numerator by dividing it by its maximal value. Then, we can wonder why Kuncheva's similarity measure sim_K (defined only for $k_1 = k_2$) belongs to $[-1,1]$ whereas Wald's measure (defined for distinct values of k_1 and k_2) does not. In fact, it can be shown that when $k_1 = k_2$, the maximal absolute value of the numerator is equal to its maximal value, so that Kuncheva's measure can be rewritten as in Theorem 2. This also proves that our proposed measure sim_N is a true generalization of Kuncheva's index as they have the same formal expression.

Theorem 2. *Kuncheva's similarity between two feature sets s_1 and s_2 of same cardinality can be rewritten as follows:*

$$sim_K(s_1, s_2) = \frac{r - \mathbb{E}[r]}{max(|r - \mathbb{E}[r]|)}. \tag{11}$$

The maximal absolute value of a term is equal to the maximum between the opposite of its minimum and its maximum. Therefore, sim_N can be rewritten as follows:

$$sim_N(s_1, s_2) = \frac{r - \mathbb{E}[r]}{max\left[-min(r - \mathbb{E}[r]); max(r - \mathbb{E}[r])\right]}. \tag{12}$$

The only variable in sim_N is the size of the intersection r and all other terms only depend on k_1, k_2 and n. Therefore, $min(r - \mathbb{E}[r]) = min(r) - \mathbb{E}[r]$ and $max(r - \mathbb{E}[r]) = max(r) - \mathbb{E}[r]$, which gives us the following expression for sim_N:

$$sim_N(s_1, s_2) = \frac{r - \mathbb{E}[r]}{max\left[-min(r) + \mathbb{E}[r]; max(r) - \mathbb{E}[r]\right]}. \tag{13}$$

As explained in Sect. 3.1, the minimal value of r is $min(r) = max(0, k_1 + k_2 - n)$ and its maximal value is $max(r) = min(k_1, k_2)$. Therefore, we have that:

$$
\begin{aligned}
sim_N(s_1, s_2) &= \frac{r - \mathbb{E}[r]}{max\left[-max(0, k_1 + k_2 - n) + \mathbb{E}[r]; min(k_1, k_2) - \mathbb{E}[r]\right]} \\
&= \frac{r - \frac{k_1 k_2}{n}}{max\left[-max(0, k_1 + k_2 - n) + \frac{k_1 k_2}{n}; min(k_1, k_2) - \frac{k_1 k_2}{n}\right]}.
\end{aligned}
\tag{14}
$$

The resulting stability measure $\overline{sim_N}$ is then taken as the average pairwise similarities as in Eq. 1. Let us now look at the properties of the new measure sim_N.

As explained previously, this measure is a proper generalization of Kuncheva's measure sim_K as it matches its value for $k_1 = k_2 = k$. By construction, this measure will be bounded by the constants -1 and 1 and reach its maximal value of 1 when the two feature sets are identical. Hence it has the Property 1 of Table 1. As outlined in the toy examples of Sect. 3.2, this allows the comparison of stability values for algorithms returning different number of features and for different values of n. It also has the Property 2 of monotonicity (as it is an increasing function of the size of the intersection r between two feature sets) and the Property 3 of correction for chance. It is invariant to feature set permutations (as it is symmetric). Figure 1 shows the maximum and minimal values of the measure in different settings. As we can see, even though this measure accounts for some of the set size variations, the maximum value is not proportional to the distance between the two subset sizes k_1 and k_2. Finally, this measure is the only one having Properties 1 to 5. As Kuncheva's similarity measure sim_K (as well as the measures sim_L, sim_W and $nPOG$), the expression of this similarity measure only holds for values of k_1 and k_2 in $\{1, ..., n-1\}$. For completeness, we will also set the values of sim_N to 0 when k_1 or k_2 is equal to 0 or n.

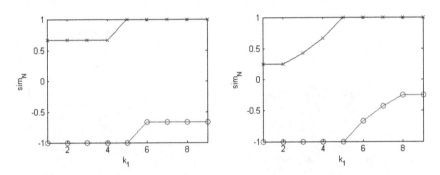

Fig. 1. Maximum and minimum of sim_N against k_1 for $k_2 = 6$ (LEFT) and $k_2 = 8$ (RIGHT) when $n = 10$.

5   Application to Feature Selection by Random Forests

To illustrate the utility of the proposed measure, we used random forests [2] as a feature selection procedure where a feature is selected when it is used in at least a percentage p of the trees. We built random forests of 100 decision trees using the mutual information as a splitting criterion. Each decision tree is built on a bootstrap sample of the given dataset. At each splitting point, the decision tree was given the choice between $\lfloor \sqrt{n} \rfloor$ features selected at random. As p is effectively a regularization parameter on the number of features selected, we tuned p for the different datasets so that only a certain proportion of the features is selected. In order to model the data perturbations, either bootstrap samples or random sub-samples can be taken [5].

Table 2. Parameters of 4 datasets, where p is the regularization parameter and where the average number of features selected using parameter p is given along with its standard deviation.

	Num. ex.	Num. classes	p	Num. feat.	Av. num. feat. selected
Wine	178	3	0.5	13	9.8 ± 0.93
Parkinsons	195	2	0.5	22	11 ± 1.7
Breast	569	2	0.5	30	15 ± 1.3
Sonar	208	2	0.25	60	42 ± 2.4

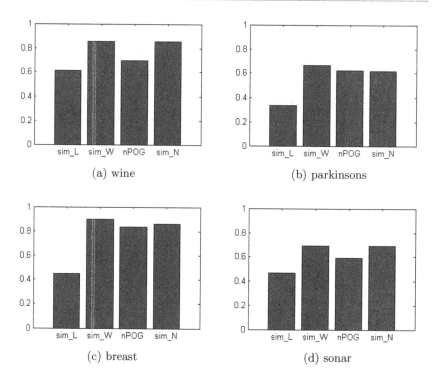

(a) wine   (b) parkinsons

(c) breast   (d) sonar

Fig. 2. Stability values on 4 datasets using the different similarity values.

Here we used $K = 100$ random sub-samples without replacement of the datasets containing 90 % of the total amount of examples [8]. So we built K random forests on each one of these samples and calculated the stability of the sequence of the K feature sets obtained. We used 4 datasets of the UCI repository, for which the properties, the values chosen for p and the average number of features selected are given in Table 2. Figure 2 gives us the stability values when using the different similarity measures. We observed that on all the datasets, the lowest stability value is obtained when using Lustgarten's similarity measure sim_L. This probably comes from the fact that sim_L does not always reach its maximal value when two feature sets are identical and its maximal value depends on the size of the feature sets selected (as observed in Toy Example 1 of Sect. 3.2). The stability value when using $nPOG$ seems closer to the value of the ones using sim_N and sim_W on the parkinsons and on the breast datasets than in the other two datasets. As we have seen in Toy Example 3, the value of $nPOG$ changes when we permute the feature sets, which makes it difficult to interpret. On the four datasets, the stability values obtained using sim_W and sim_N are close to each other. This can be explained by the fact that in some situations (i.e. for certain values of k_1, k_2 and n), the value sim_N will be equal to the one of sim_W. Indeed, when we take a pair of feature sets s_1 and s_2, if we have k_1, k_2 and n such that $-min(r) + \mathbb{E}[r] \leq max(r) - \mathbb{E}[r]$, the denominator of the two similarity measures becomes the same and in that case $sim_W(s_1, s_2) = sim_N(s_1, s_2)$. In other words, in the feature sets returned by this procedure, only a small proportion of pairs of feature sets do not satisfy this. We have seen in Toy Example 4 that the minimal value of sim_W decreases with n and this could strongly bias the resulting stability value in some cases. This situation happens when the feature sets are very dissimilar in both terms of cardinality and of features selected. In the four datasets, we can observe that this is not the case as the standard deviations of the number of features selected by the random forests are much smaller than the total number of features.

6   Conclusion

Through a thorough study of the literature, we identified a set of desirable properties for stability measures dealing with feature selection procedures that return feature sets. After leading a comparative study on the measures that have the property of correction for chance, a generalization of Kuncheva's index is proposed for feature selection algorithms that do not return feature sets of constant cardinality. This new measure has all the desired properties except that it does not take into account possible redundancy between features, which could be the focus of future work. We illustrate a possible application of this measure in the context of ensemble-based feature selection and exhibit the differences obtained in the stability values using the different measures.

Acknowledgements. This work was supported by the EPSRC grant [EP/I028099/1].

References

1. Abeel, T., Helleputte, T., Van de Peer, Y., Dupont, P., Saeys, Y.: Robust biomarker identification for cancer diagnosis with ensemble feature selection methods. Bioinformatics **26**, 392–398 (2010)
2. Breiman, L.: Random forests. Mach. Learn. **45**, 5–32 (2001)
3. Ditzler, G., Polikar, R., Rosen, G.: A bootstrap based neyman-pearson test for identifying variable importance. IEEE Trans. Neural Netw. Learn. Syst. **26**, 880–886 (2014)

4. Dunne, K., Cunningham, P., Azuaje, F.: Solutions to instability problems with sequential wrapper-based approaches to feature selection. Technical report, Journal of Machine Learning Research (2002)
5. He, Z., Yu, W.: Stable feature selection for biomarker discovery. Comput. Biol. Chem. **34**, 215–225 (2010)
6. Kalousis, A., Prados, J., Hilario, M.: Stability of feature selection algorithms: a study on high-dimensional spaces. Knowl. Inf. Syst. **12**, 95–116 (2007)
7. Křížek, P., Kittler, J., Hlaváč, V.: Improving stability of feature selection methods. In: Kropatsch, W.G., Kampel, M., Hanbury, A. (eds.) CAIP 2007. LNCS, vol. 4673, pp. 929–936. Springer, Heidelberg (2007)
8. Kuncheva, L.I.: A stability index for feature selection. In: Artificial Intelligence and Applications (2007)
9. Lustgarten, J.L., Gopalakrishnan, V., Visweswaran, S.: Measuring stability of feature selection in biomedical datasets. In: Proceedings of the AMIA Annual Symposium (2009)
10. Saeys, Y., Abeel, T., Van de Peer, Y.: Robust feature selection using ensemble feature selection techniques. In: Daelemans, W., Goethals, B., Morik, K. (eds.) ECML PKDD 2008, Part II. LNCS (LNAI), vol. 5212, pp. 313–325. Springer, Heidelberg (2008)
11. Somol, P., Novovičová, J.: Evaluating stability and comparing output of feature selectors that optimize feature subset cardinality. IEEE Trans. Pattern Anal. Mach. Intell. **32**, 1921–1939 (2010)
12. Wald, R., Khoshgoftaar, T.M., Napolitano, A.: Stability of filter- and wrapper-based feature subset selection. In: International Conference on Tools with Artificial Intelligence. IEEE Computer Society (2013)
13. Yu, L., Ding, C.H.Q., Loscalzo, S.: Stable feature selection via dense feature groups. In: KDD (2008)
14. Zhang, M., Zhang, L., Zou, J., Yao, C., Xiao, H., Liu, Q., Wang, J., Wang, D., Wang, C., Guo, Z.: Evaluating reproducibility of differential expression discoveries in microarray studies by considering correlated molecular changes. Bioinformatics **25**, 1662–1668 (2009)

Suboptimal Graph Edit Distance Based on Sorted Local Assignments

Kaspar Riesen[1,2]([⊠]), Miquel Ferrer[1], and Horst Bunke[2]

[1] Institute for Information Systems, University of Applied Sciences and Arts
Northwestern Switzerland, Riggenbachstrasse 16, 4600 Olten, Switzerland
{kaspar.riesen,miquel.ferrer}@fhnw.ch
[2] Institute of Computer Science and Applied Mathematics, University of Bern,
Neubrückstrasse 10, 3012 Bern, Switzerland
{riesen,bunke}@iam.unibe.ch

Abstract. Graph based pattern representation offers a number of useful properties. In particular, graphs can adapt their size and complexity to the actual pattern, and moreover, graphs are able to describe structural relations that might exist within a pattern. Yet, the high representational power and flexibility of graphs is accompanied by a significant increase of the complexity of many algorithms. For instance, exact computation of pairwise graph distance can be accomplished in exponential time complexity only. A previously introduced approximation framework reduces the problem of graph distance computation to an instance of a linear sum assignment problem. This allows suboptimal graph distance computation in cubic time. The present paper introduces a novel procedure, which is conceptually related to this previous approach, but offers $O(n^2 \log(n^2))$ (rather than cubic) run time. We empirically verify the speed up of our novel approximation and show that the faster approximation is able to keep up with the existing framework with respect to distance accuracy.

1 Introduction

In pattern recognition applications where the underlying data is characterized by complex structural relationships (e.g. [1–3]), graphs are often used as basic formalism for pattern representation. The process of evaluating the dissimilarity of graphs is referred to as graph matching [4,5]. Among a vast number of graph matching methods available [6–8], the concept of graph edit distance [9,10] is in particular interesting because it is able to cope with directed and undirected, as well as with labeled and unlabeled graphs. If there are labels on nodes, edges, or both, no constraints on the respective label alphabets have to be considered. In fact, graph edit distance is a widely accepted concept for general graph dissimilarity computation and has been used in various applications [11–13].

A well known drawback of graph edit distance is its computational complexity which is exponential in the number of nodes of the involved graphs. This means that for large graphs the exact computation of graph edit distance is intractable. In recent years, a number of methods addressing the high complexity of graph

© Springer International Publishing Switzerland 2015
F. Schwenker et al. (Eds.): MCS 2015, LNCS 9132, pp. 147–156, 2015.
DOI: 10.1007/978-3-319-20248-8_13

edit distance computation have been proposed [14,15]. In [16] the authors of the present paper introduced an algorithmic framework that reduces the problem of graph edit distance computation to a *linear sum assignment problem* (LSAP). The basic idea is to subdivide the graphs into individual nodes including local structural information. Next, an LSAP solving algorithm is employed in order to find an optimal assignment of the nodes (plus local structures) of both graphs in cubic time. Finally, an approximate graph edit distance, which is globally consistent with the underlying edge structures of both graphs, is derived from the assignment of local substructures.

The major goal of the present paper is to speed up the approximation framework presented in [16]. In particular, we aim at substantially speeding up the assignment of local substructures. To this end we replace the optimal assignment algorithm with cubic time complexity with a suboptimal algorithm which runs in linearithmic time. The basic idea of our novel approach is to sort all possible assignments with respect to their individual local costs in a list in ascending order. Next, the nodes of both graphs are unambiguously assigned to each other in a greedy manner according to the assignments that occur first in the list of costs from head to tail.

The remainder of the present paper is organized as follows. Next, in Sect. 2 the concept of graph edit distance and the original framework for graph edit distance approximation [16] is summarized. In Sect. 3 the novel assignment algorithm is introduced. An experimental evaluation on diverse data sets is carried out in Sect. 4, and in Sect. 5 we draw conclusions and point out possible directions for future work.

2   Bipartite Graph Edit Distance Approximation

2.1   Graph Edit Distance

A graph g is a four-tuple $g = (V, E, \mu, \nu)$, where V is the finite set of nodes, $E \subseteq V \times V$ is the set of edges, $\mu : V \to L_V$ is the node labeling function, and $\nu : E \to L_E$ is the edge labeling function. The labels for both nodes and edges can be given by the set of integers $L = \{1, 2, 3, \ldots\}$, the vector space $L = \mathbb{R}^n$, a set of symbolic labels $L = \{\alpha, \beta, \gamma, \ldots\}$, or a combination of various label alphabets from different domains. Unlabeled graphs are obtained as a special case by assigning the same (empty) label $\varnothing$ to all nodes and edges.

Given two graphs, the source graph $g_1 = (V_1, E_1, \mu_1, \nu_1)$ and the target graph $g_2 = (V_2, E_2, \mu_2, \nu_2)$, the basic idea of graph edit distance [9,10] is to transform g_1 into g_2 using some edit operations. A standard set of edit operations is given by *insertions*, *deletions*, and *substitutions* of both nodes and edges. We denote the substitution of two nodes $u \in V_1$ and $v \in V_2$ by $(u \to v)$, the deletion of node $u \in V_1$ by $(u \to \varepsilon)$, and the insertion of node $v \in V_2$ by $(\varepsilon \to v)$, where ε refers to the empty node. For edge edit operations we use a similar notation.

A sequence $(e_1, \ldots, e_k)$ of k edit operations e_i that transform g_1 completely into g_2 is called *edit path* $\lambda(g_1, g_2)$ between g_1 and g_2. Let $\Upsilon(g_1, g_2)$ denote the set of all admissible and complete edit paths between two graphs g_1 and g_2.

To find the most suitable edit path out of $\Upsilon(g_1, g_2)$, one introduces a cost $c(e)$ for every edit operation e, measuring the strength of the corresponding operation. The idea of such a cost is to define whether or not an edit operation represents a strong modification of the graph. By means of cost functions for elementary edit operations, graph edit distance allows the integration of domain specific knowledge about object similarity. Furthermore, if in a particular case prior knowledge about the labels and their meaning is not available, automatic procedures for learning the edit costs from a set of sample graphs are available as well [17].

Clearly, between two similar graphs, there should exist an inexpensive edit path, representing low cost operations, while for dissimilar graphs an edit path with high cost is needed. Consequently, the exact edit distance $d_{\lambda_{\min}}(g_1, g_2)$ of two graphs g_1 and g_2 is defined as the sum of cost of the minimal cost edit path found in $\Upsilon(g_1, g_2)$.

2.2 Approximation of Graph Edit Distance

Algorithms for computing the exact edit distance $d_{\lambda_{\min}}$ are typically based on combinatorial search procedures. The search space for those procedures is $\Upsilon(g_1, g_2)$, which contains $O(m^n)$ edit paths to be explored (assuming m nodes in g_1 and n nodes in g_2). Hence, the computational complexity of exact graph edit distance is exponential in the number of nodes of the involved graphs. This means that for large graphs the computation of edit distance is intractable.

The graph edit distance approximation framework introduced in [16] reduces the difficult problem of graph edit distance computation to an instance of a *Linear Sum Assignment Problem (LSAP)* for which a large number of efficient algorithms exist [18].

By reformulating graph edit distance as an instance of an LSAP (denoted with $BP\text{-}GED$[1]), three major steps have to be carried out.

First Step: A square *cost matrix* $\mathbf{C} = (c_{ij})$ based on the node sets $V_1 = \{u_1, \ldots, u_n\}$ and $V_2 = \{v_1, \ldots, v_m\}$ of g_1 and g_2, respectively, is established as follows.

$$\mathbf{C} = \begin{bmatrix} c_{11} & c_{12} & \cdots & c_{1m} & c_{1\varepsilon} & \infty & \cdots & \infty \\ c_{21} & c_{22} & \cdots & c_{2m} & \infty & c_{2\varepsilon} & \ddots & \vdots \\ \vdots & \vdots & \ddots & \vdots & \vdots & \ddots & \ddots & \infty \\ c_{n1} & c_{n2} & \cdots & c_{nm} & \infty & \cdots & \infty & c_{n\varepsilon} \\ c_{\varepsilon 1} & \infty & \cdots & \infty & 0 & 0 & \cdots & 0 \\ \infty & c_{\varepsilon 2} & \ddots & \vdots & 0 & 0 & \ddots & \vdots \\ \vdots & \ddots & \ddots & \infty & \vdots & \ddots & \ddots & 0 \\ \infty & \cdots & \infty & c_{\varepsilon m} & 0 & \cdots & 0 & 0 \end{bmatrix} \quad (1)$$

---

[1] *Bipartite Graph Edit Distance* (LSAPs can be formulated by means of *bipartite graphs*).

Entry c_{ij} thereby denotes the cost of a node substitution $(u_i \rightarrow v_j)$, $c_{i\varepsilon}$ denotes the cost of a node deletion $(u_i \rightarrow \varepsilon)$, and $c_{\varepsilon j}$ denotes the cost of a node insertion $(\varepsilon \rightarrow v_j)$. That is, the left upper corner of $\mathbf{C} = (c_{ij})$ represents the costs of all possible node substitutions, while the diagonals of the right upper and left lower corner represent the costs of all possible node deletions and insertions, respectively (every node can be deleted or inserted at most once and thus any non-diagonal element can be set to ∞ in these parts). Substitutions of the form $(\varepsilon \rightarrow \varepsilon)$ should not cause any cost and therefore, any element in the right lower part is set to zero.

Second Step: Next, an LSAP is stated on cost matrix $\mathbf{C} = (c_{ij})$ and eventually solved. The LSAP optimization consists in finding a permutation $(\varphi_1, \ldots, \varphi_{n+m})$ of the integers $(1, 2, \ldots, (n + m))$ that minimizes the overall assignment cost $\sum_{i=1}^{(n+m)} c_{i\varphi_i}$. This permutation corresponds to the assignment

$$\psi = ((u_1 \rightarrow v_{\varphi_1}), (u_2 \rightarrow v_{\varphi_2}), \ldots, (u_{m+n} \rightarrow v_{\varphi_{m+n}}))$$

of the nodes of g_1 to the nodes of g_2. Note that assignment ψ includes node assignments of the form $(u_i \rightarrow v_j)$, $(u_i \rightarrow \varepsilon)$, $(\varepsilon \rightarrow v_j)$, and $(\varepsilon \rightarrow \varepsilon)$ (the latter can be dismissed, of course). Hence, the definition of $\mathbf{C} = (c_{ij})$ in Eq. 1 explicitly allows insertions and/or deletions to occur in an optimal assignment.

The optimal assignment ψ does not take any structural constraints on the graphs into account as long as the individual entries in $\mathbf{C} = (c_{ij})$ consider the nodes of both graphs only. In order to integrate knowledge about the graph structure, to each entry c_{ij}, i.e. to each cost of a node edit operation $(u_i \rightarrow v_j)$, the minimum sum of edge edit operation costs, implied by the corresponding node operation, is added. This enables the LSAP to consider information about the local, yet not global, edge structure of a graph for optimizing the node assignment.

Third Step: The LSAP optimization finds an assignment ψ in which every node of g_1 and g_2 is either assigned to a unique node of the other graph, deleted or inserted. Note that edit operations on edges are always defined by the edit operations on their adjacent nodes. That is, whether an edge (u, v) is substituted, deleted, or inserted, depends on the edit operations actually performed on both adjacent nodes u and v. Since ψ refers to a consistent and complete node assignment, the edge operations can be completely and consistently inferred from ψ. Hence, we get a valid edit path $\lambda_\psi \in \Upsilon(g_1, g_2)$ between the graphs under consideration.

Yet, λ_ψ considers the edge structure of g_1 and g_2 in a global and consistent way, while the underlying optimal node assignment ψ is able to consider the structural information in an isolated way only (single nodes and their adjacent edges). Hence, the edit path λ_ψ found by this specific framework is not necessarily optimal and thus the corresponding distance d_ψ is – in the best case – equal to, or – in general – larger than the exact graph edit distance $d_{\lambda_{\min}}$.

3   Sort Match for Graph Edit Distance Approximation

The prime reason for building a square cost matrix $\mathbf{C} = (c_{ij})$ in [16] is to formulate a standard LSAP that takes the peculiarities of graph edit distance into account. This particular LSAP (defined on $\mathbf{C} = (c_{ij})$) is eventually solved in an optimal manner. The novel method of the present paper is similar to [16] in the sense of first regarding the individual node sets $V_1 = \{u_1, \ldots, u_n\}$ and $V_2 = \{v_1, \ldots, v_m\}$ of the involved graphs only. Yet, in contrast with [16] where a square cost matrix is built upon V_1 and V_2, our novel procedure is based on a list of costs established as follows.

$$\mathbf{l} = \{\underbrace{c_{11}, c_{12}, \ldots c_{ij}, \ldots, c_{nm}}_{\text{substitutions}}, \underbrace{c_{1\varepsilon}, c_{2\varepsilon}, \ldots, c_{n\varepsilon}}_{\text{deletions}}, \underbrace{c_{\varepsilon 1}, c_{\varepsilon 2}, \ldots, c_{\varepsilon m}}_{\text{insertions}}\} \qquad (2)$$

Obviously, list $\mathbf{l}$ buffers the $n \times m$ elements that represent the costs of all possible node substitutions (left upper corner of $\mathbf{C}$), as well as n and m elements that represent the costs of all possible node deletions and insertions, respectively (diagonals of the right upper and left lower corner of $\mathbf{C}$). Although list $\mathbf{l}$ contains less entries than $\mathbf{C}$, it essentially contains the same information as $\mathbf{C}$ as the omitted elements are ∞- and 0-elements only. Yet, major benefit of our novel procedure is not the downsized number of elements to be considered, but a suboptimal – rather than an optimal – assignment algorithm based on local assignment costs.

The complete approximation framework is outlined in Algorithm 1. First, the basic list $\mathbf{l}$ is built and eventually sorted in ascending order using an optimized merge sort in $O(n \log n)$ time [19] (line 1 and 2). On line 3 and 4, ψ is initialized as empty set of assignments, counter k is set to 1 and all nodes of both graphs are marked as available. Next, as long as not all nodes of g_1 and g_2 are processed, the individual cost entries c_{ij} are visited from head to tail (line 5 and 6). On line 7 it is verified whether the pair of indices (i, j) (corresponding to the currently visited entry c_{ij}) is admissible. Remember that every element c_{ij} uniquely corresponds to a certain node edit operation $(u_i \to v_j)$. A pair of indices (i, j) is admissible if one of the following three cases is true:

1. both nodes $u_i \in V_1, v_j \in V2$ are available (substitution of unprocessed nodes u_i, v_j)
2. $i == \varepsilon$ and $v_j \in V_2$ is available (insertion of unprocessed node v_j)
3. $j == \varepsilon$ and $u_i \in V_1$ is available (deletion of unprocessed node u_i)

If (i, j) is admissible, the node edit operation $(u_i \to v_j)$ is added to the set of assignments ψ in the basic version of the algorithm (line 13). Eventually, the corresponding nodes u_i and v_j are marked as unavailable[2] (line 14).

In a refined version of our procedure, the subroutine *Look-Ahead* (Algorithm 2) with user defined parameter δ is called, if (i, j) is admissible (line 8).

---

[2] The i-th node $u_i \in V_1$ is marked as unavailable only, if the corresponding index i is not equal to ε, of course. The same accounts for index j and node $v_j \in V_2$.

Algorithm 1. Sort-Match(g_1, g_2, δ)

1: Build list $l = \{c_{11}, \ldots, c_{\varepsilon m}\}$ according to Eq. 2 and input graphs g_1, g_2
2: Sort elements $c_{ij} \in l$ in ascending order: $l = (c^{(1)}, \ldots, c^{(nm+n+m)})$
3: $\psi = \{\}$
4: $k = 1$; mark all nodes in g_1 and g_2 as available
5: **while** nodes in g_1 are available **or** nodes in g_2 are available **do**
6:    Let c_{ij} be the k-th element $c^{(k)}$ in list l
7:    **if** (i, j) is *admissible* **then**
8:       $\tau = Look\text{-}Ahead(\delta)$
9:       **if** $|\tau| > 0$ **then**
10:          $\psi = \psi \cup \tau$
11:          Mark $u_i, u_{i'}, v_j$ and $v_{j'}$ as unavailable
12:       **else**
13:          $\psi = \psi \cup \{(u_i \rightarrow v_j)\}$
14:          Mark u_i and v_j as unavailable
15:       **end if**
16:    **end if**
17:    $k = k + 1$
18: **end while**
19: Complete edit path according to ψ and **return** $d_\psi(g_1, g_2)$

Algorithm 2. Look-Ahead(δ)

1: $\delta = \min(\delta, (nm + n + m - k))$
2: Assignments $A = \{(u_{(1)} \rightarrow v_{(1)}), \ldots, (u_{(\delta)} \rightarrow v_{(\delta)})\}$ correspond to $\{c^{(k+1)}, \ldots, c^{(k+\delta)}\}$
3: **if** $\exists \{(u_i \rightarrow v_{j'}), (u_{i'} \rightarrow v_j)\} \subseteq A$ with admissible pairs (i, j'), (i', j) **then**
4:    **return** $\{(u_i \rightarrow v_{j'}), (u_{i'} \rightarrow v_j)\}$ **else return** $\{\}$
5: **end if**

This subroutine browses through the subsequent δ elements $\{c^{(k+1)}, \ldots, c^{(k+\delta)}\}$ of the currently processed element (note that line 1 of Algorithm 2 ensures that the search area remains in list l even if the number of unvisited elements in l is smaller than δ). Overall aim of the subroutine is to find two alternative admissible assignments $(u_i \rightarrow v_{j'})$ and $(u_{i'} \rightarrow v_j)$ for both nodes $u_i \in V_1$ and $v_j \in V_2$ within the next δ elements in l. If we find such a pair, we return it to the calling algorithm and save it in τ on line 8 (otherwise an empty set is returned and saved in τ). If τ is not empty, the assignments in τ are added to ψ and the corresponding nodes are marked as unavailable (line 10 and 11). Otherwise the initial admissible edit operation is added to ψ (line 13 and 14).

The intuition behind this refinement is as follows. The basic assignment process works in a greedy manner. That is, once an admissible edit operation $(u_i \rightarrow v_j)$ has been found in l, it is irrevocably added to ψ. However, this particular edit operation might prevent two assignments that involve indices i and j which would have been processed in the next δ steps. Clearly, as list l is in ascending order, the individual entries $c_{ij'}$ and $c_{i'j}$ have to be greater than or equal to c_{ij}. Yet, in contrast with $(u_i \rightarrow v_j)$, which includes at most two nodes, the combination of the two alternative assignments include four nodes in the best case. Hence, though the individual cost entries are greater than (or equal to) the current admissible assignment, it might be reasonable to add $\{(u_i \rightarrow v_{j'}), (u_{i'} \rightarrow v_j)\}$ to ψ rather than only $(u_i \rightarrow v_j)$.

Similar to BP-GED, the proposed algorithmic procedure finds an assignment ψ in which every node of g_1 and g_2 is either assigned to a unique node of the other graph, deleted or inserted. Hence, on line 19 we are able to infer all necessary edge edit operations to get a complete edit path $\lambda_\psi \in \Upsilon(g_1, g_2)$ and the corresponding approximate edit distance d_ψ. For the remainder of this paper we denote this adapted algorithmic procedure with *Sort-Match-Look-Ahead(δ)-GED* (or *SMLA(δ)-GED* for short).

In contrast with the optimal assignment ψ, the node assignment ψ' of our novel procedure is suboptimal. That is, the sum of assignments costs of our novel approach is greater than, or equal to, the minimal assignment cost provided by optimal LSAP solving algorithms. However, note that for the corresponding distance values d_ψ and $d_{\psi'}$ no globally valid order relation exists. That is, the approximate graph edit distance $d_{\psi'}$ derived from ψ' can be greater than, equal to, or smaller than d_ψ.

The time critical part of both algorithms BP-GED and SMLA(δ)-GED is the process of finding the assignment of local structures. BP-GED defines the assignment by solving the LSAP on $\mathbf{C} = (c_{ij})$ in an optimal way which takes $O((n+m)^3)$ time. Clearly, the bottleneck of the assignment process in SMLA(δ)-GED is the sorting of list l which can be accomplished in linearithmic time with respect to the number of elements in l. Assuming $n \approx m$, we have approximately $n^2 + 2n$ elements in l and thus the overall complexity of graph edit distance computation amounts to $O(n^2 \log(n^2))$.

4 Experimental Evaluation

The goal of the experimental evaluation, carried out on three different data sets from the IAM graph repository [20], is twofold. First, we aim at empirically confirming the faster matching time of SMLA(δ)-GED compared to BP-GED. Second, we aim at answering the question whether or not the SMLA graph edit distances remain sufficiently accurate for graph based pattern classification.

In our experimental evaluation we use four different values for δ, viz. $\{0, 5, 10, 15\}$. For every graph pair we incrementally increase the size of the look ahead area from 0 to δ with step size 5. Eventually, we return the minimum edit distance found by all approximation variants. Hence, for a specific graph pair (g_1, g_2) and a certain value of $\delta \in \{0, 5, 10, 15\}$, we have

$$\text{SMLA}(\delta)\text{-GED}(g_1, g_2) = \min_{i=0,5,\ldots,\delta} \{\text{SMLA}(i)\text{-GED}(g_1, g_2)\}$$

Clearly by means of this specific experimental set up, the distances between all pairs of graphs monotonically decrease when parameter δ is increased. This can actually be observed in Table 1 on the lines that show the results for $\varnothing e$. Characteristic number $\varnothing e$ measures the mean relative over- and underestimation of SMLA graph edit distances compared with BP-GED[3]. We observe that

---

[3] Note that both means are computed on the sets of distances where an SMLA approach actually over- or underestimates the original approximation.

Table 1. The mean relative deviation of SMLA-GED algorithm variants compared with BP-GED in percentage ($\varnothing e$), the accuracy of a 1NN classifier in percentage, and the mean run time for one matching in ms ($\varnothing t$).

Data		BP-GED	SMLA(0)-GED	SMLA(5)-GED	SMLA(10)-GED	SMLA(15)-GED
				Algorithm		
AIDS	$\varnothing e$ [%]	–	+5.3/ − 4.4	+5.1/ − 4.9	+4.9/ − 5.2	+4.8/ − 5.3
	1NN [%]	99.07	99.27	99.27	99.20	99.20
	$\varnothing t$ [ms]	3.61	1.24	1.35	1.43	1.51
FP	$\varnothing e$ [%]	–	+5.7/ − 18.2	+5.3/ − 18.5	+5.1/ − 18.7	+5.0/ − 18.8
	1NN [%]	79.75	75.95	76.50	77.20	77.50
	$\varnothing t$ [ms]	0.41	0.32	0.32	0.32	0.33
MUTA	$\varnothing e$ [%]	–	+5.8/ − 4.9	+5.7/ − 5.0	+5.6/ − 5.2	+5.5/ − 5.2
	1NN [%]	70.20	72.50	72.60	72.50	72.10
	$\varnothing t$ [ms]	33.89	4.63	5.28	5.72	6.04

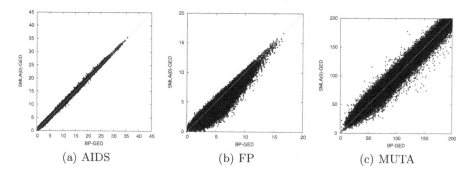

   (a) AIDS   (b) FP   (c) MUTA

Fig. 1. Distances of BP-GED (x-axis) vs. distances of SMLA(0)-GED (y-axis).

increased values of δ reduce the mean overestimation on all data sets while the mean underestimation is further increased. On the AIDS and MUTA data sets the means of over- and underestimations are quite balanced. Yet, on the FP data set the overestimation is accompanied with a heavy underestimation. The over- and underestimation of SMLA based edit distance approximation compared with BP-GED (and in particular the heavy underestimation of several distances on the FP data) can also be observed in the scatter plots in Fig. 1.

Next, Table 1 shows the recognition rate of a 1-nearest-neighbor classifier (1NN). The nearest neighbor paradigm is particularly interesting for the present evaluation because it directly uses the distances without any additional classifier training. In comparison with BP-GED we observe that SMLA-GED deteriorates the recognition rates on the FP data set (all deteriorations are statistically significant ($\alpha = 0.05$)). That is, the heavy underestimation observed above on this data set, crucially disturbs the nearest neighbor classification. However, on the other two data sets SMLA(δ)-GED outperforms BP-GED wit any value of δ (the first three improvements on the MUTA data set are statistically significant).

Finally we focus on the mean run time for one matching in ms ($\varnothing t$). On the relatively small graphs of the FP data set, the speed-up by our novel approximation compared to BP-GED is rather small. Yet, on the other two data sets with larger graphs substantial speed ups can be observed. That is, using SMLA(0)-GED rather than BP-GED on the AIDS data set leads to a decrease of the mean matching time from 3.61 ms to 1.24 ms. On the MUTA data SMLA(0)-GED is more than seven times faster than the original approximation. We also observe that increasing the size of the look ahead window δ from 0 to 15 moderately increases the run time on all data sets (as expected). Yet, the run time of SMLA(δ)-GED remains below run time of BP-GED for any tested value of δ.

5 Conclusions and Future Work

The present paper proposes a novel graph edit distance approximation algorithm. The main idea is to build a list which comprehends the individual costs for all possible substitutions, deletions, and insertions of local graph structures. Next, we sort the list, iterate trough it from head to tail, and incrementally add admissible edit operations to the edit path (until all nodes of both graphs are processed). We additionally implement a look ahead procedure which verifies whether the current operation possibly prevents other beneficial edit operations that occur in the next steps of the list search. The novel algorithm allows graph edit distance approximation in $O(n^2 \log(n^2))$ time.

The speed up of our novel approximation compared to a previous approximation algorithm is empirically verified on three different graph data sets. Moreover, we observe that in two out of three cases the proposed approximation algorithms are able to keep up with, or even surpass, the existing framework with respect to recognition accuracy using a 1NN classifier. In future work we aim at testing our novel approach on additional graph data sets and especially larger graphs.

Acknowledgements. This work has been supported by the *Hasler Foundation* Switzerland and the *Swiss National Science Foundation* project 200021_153249.

References

1. Lumini, A., Maio, D., Maltoni, D.: Inexact graph matching for fingerprint classification. Mach. Graph. Vis. Spec. Issue Gr. Transform. Pattern Gener. CAD **8**(2), 231–248 (1999)
2. Richiardi, J., Achard, S., Bunke, H., Van De Ville, D.: Machine learning with brain graphs: predictive modeling approaches for functional imaging in systems neuroscience. IEEE Signal Process. Mag. **30**(3), 58–70 (2013)
3. Cesare, S., Xiang, Y.: Malware variant detection using similarity search over sets of control flow graphs. In: Proceedings of 10th International Conference on Trust, Security and Privacy in Computing and Communications, pp. 181–189 (2011)

4. Conte, D., Foggia, P., Sansone, C., Vento, M.: Thirty years of graph matching in pattern recognition. Int. J. Pattern Recog. Artif. Intell. **18**(3), 265–298 (2004)
5. Foggia, P., Percannella, G., Vento, M.: Graph matching and learning in pattern recognition in the last 10 years. Int. J. Pattern Recog. Art. Intell. Online Ready **28** (2014)
6. Gaüzère, B., Brun, L., Villemin, D.: Two new graphs kernels in chemoinformatics. Pattern Recogn. Lett. **33**(15), 2038–2047 (2012)
7. Rossi, L., Torsello, A., Hancock, E.R.: A continuous-time quantum walk kernel for unattributed graphs. In: Kropatsch, W.G., Artner, N.M., Haxhimusa, Y., Jiang, X. (eds.) GbRPR 2013. LNCS, vol. 7877, pp. 101–110. Springer, Heidelberg (2013)
8. Emms, D., Hancock, E.R., Wilson, R.C.: A correspondence measure for graph matching using the discrete quantum walk. In: Escolano, F., Vento, M. (eds.) GbRPR. LNCS, vol. 4538, pp. 81–91. Springer, Heidelberg (2007)
9. Bunke, H., Allermann, G.: Inexact graph matching for structural pattern recognition. Pattern Recogn. Lett. **1**, 245–253 (1983)
10. Sanfeliu, A., Fu, K.: A distance measure between attributed relational graphs for pattern recognition. IEEE Trans. Syst. Man Cybern. (Part B) **13**(3), 353–363 (1983)
11. Robles-Kelly, A., Hancock, E.: Graph edit distance from spectral seriation. IEEE Trans. Pattern Anal. Mach. Intell. **27**(3), 365–378 (2005)
12. Myers, R., Wilson, R., Hancock, E.: Bayesian graph edit distance. IEEE Trans. Pattern Anal. Mach. Intell. **22**(6), 628–635 (2000)
13. Rebagliati, N., Solé-Ribalta, A., Pelillo, M., Serratosa, F.: Computing the graph edit distance using dominant sets. In: Proceedings of 21st International Conference on Pattern Recognition, pp. 1080–1083 (2012)
14. Sorlin, S., Solnon, C.: Reactive tabu search for measuring graph similarity. In: Brun, L., Vento, M. (eds.) GbRPR 2005. LNCS, vol. 3434, pp. 172–182. Springer, Heidelberg (2005)
15. Justice, D., Hero, A.: A binary linear programming formulation of the graph edit distance. IEEE Trans. Pattern Anal. Mach. Intell. **28**(8), 1200–1214 (2006)
16. Riesen, K., Bunke, H.: Approximate graph edit distance computation by means of bipartite graph matching. Image Vis. Comput. **27**(4), 950–959 (2009)
17. Caetano, T., McAuley, J.J., Cheng, L., Le, Q., Smola, A.: Learning graph matching. IEEE Trans. Pattern Anal. Mach. Intell. **31**(6), 1048–1058 (2009)
18. Burkard, R., Dell'Amico, M., Martello, S.: Assignment Problems. Society for Industrial and Applied Mathematics, Philadelphia (2009)
19. Knuth, D.: 5.2.4: sorting by merging. In: Sorting and Searching. The Art of Computer Programming 3, pp. 158–168. Addison Wesley (1998)
20. Riesen, K., Bunke, H.: IAM graph database repository for graph based pattern recognition and machine learning. In: da Vitoria Lobo, N., Kasparis, T., Roli, F., Kwok, J.T., Georgiopoulos, M., Anagnostopoulos, G.C., Loog, M. (eds.) SSPR & SPR 2008. LNCS, vol. 5342, pp. 287–297. Springer, Heidelberg (2008)

Application and Evaluation

Multimodal PLSA for Movie Genre Classification

Hao-Zhi Hong and Jen-Ing G. Hwang[(✉)]

Department of Computer Science and Information Engineering, Fu Jen Catholic University, New Taipei City 24205, Taiwan
401226279@mail.fju.edu.tw, jihwang@csie.fju.edu.tw

Abstract. The aim of this paper is to categorize movies into genres using the previews. Our study attempts to combine audio, visual and text features to classify a collection of movie previews into action, biography, comedy, and horror. For each of the collected previews, the audio and visual features are extracted and the text features are drawn from social tags via social websites. The probabilistic latent semantic analysis (PLSA) is used to incorporate the features from these three different aspects of information. The standard PLSA processes one type of information only. Therefore double-model and triple-model PLSAs are extended in order to combine two or three different types of information. We compare these various variants of PLSA approaches with unimodal PLSAs, which use either audio, visual or text features only. The experimental results show not only that one of the triple-model PLSAs achieves the highest accuracy, but also that social tags (text features) play an important role for classifying movies genres.

Keywords: Probabilistic latent semantic analysis · Movie genres · Social tags · Audio features · Visual features

1 Introduction

Both advanced multimedia technology and communications have led the significant increase in the popularity of the digital video. The widespread use of the digital video has given people access to various digital data such as homemade videos, micro films, and movies. A great proportion of the mainstream media and the entertainment industry consists of movies. Moreover, watching movies is always a favorite form of entrainment for most people. With the high technology of multimedia and the inexpensive charge of internet access, it is much easier to watch movies online. This leads to the necessity of classifying movies into different genres to help people search for a movie of their interest.

Automatic genre classification of movies has become important with the dramatically increasing number of movies. For the purpose of movie genre classification, features are drawn from three data types: audio, visual, and text. Brezeale and Cook [1] used closed captions as text features and performed text and visual features separately in order to classify movie genres. Moncrieff et al. [2] used audio-based cinematic principles to identify with both horror and non-horror movies. Most of the approaches to movie genre classification relied on audio and visual features [3–8]. Some of the approaches incorporated cinematic principles or concepts from film theory. For example, Rasheed and Shah [9]

© Springer International Publishing Switzerland 2015
F. Schwenker et al. (Eds.): MCS 2015, LNCS 9132, pp. 159–167, 2015.
DOI: 10.1007/978-3-319-20248-8_14

used a combination of audio features, visual features and cinematic principles for genre categorization.

Automatic genre classification of movies is a challenging task. In this paper we attempt to combine audio, visual and text features for movie genre classification. With the rise of social networking sites, tags are keywords or terms supplied by online communities and provide useful information to users. This study has collected 140 movie previews from IMDB [10] and drawn their corresponding social tags from [11]. The social tags are represented as text features, while the audio and visual features are extracted from movie previews directly. To overcome the difficulty of combing three aspects of information, the standard probabilistic latent semantic analysis (PLSA) [12] is extended to fuse various types of information. Levy et al. [13] has extended the standard PLSA to integrate social tags with audio features for music information retrieval. Lienhart et al. [14] has also extended the PLSA to combine text and visual features for image retrieval. As both extensions, which have had satisfactory results, have inspired us to develop triple-model PLSAs that combine audio, visual and text features for movie genre classification.

The remainder of this paper is structured as follows. In the next section, we introduce the standard PLSA. Section 3 describes various variants of PLSAs for integrating different types of information. Section 4 displays experimental results. Finally, we provide conclusions in Sect. 5.

2   Standard PLSA

In 2001, Hofman [12] proposed the PLSA, which is a topic model based on probabilistic and statistics and is mainly used to discover the distribution of hidden topics in a text document. The term-document matrix of the PLSA is shown in Fig. 1, where the vertical axis represents M documents, the horizontal axis represents N words, and $n(d_i, v_j)$ represents the count of the j-th word appeared in the i-th document. As users are usually able to observe document and word data and therefore their relation may form a term-document matrix. The PLSA observes the relationship between the document (d) and the word (v) by means of hidden topics (z). This generates the PLSA model expressed by the following equation:

$$P(d, v) = P(d)P(v|d) = P(d) \sum_{z \in Z} P(v|z)P(z|d) \tag{1}$$

where $z \in Z = \{z_1, z_2, \ldots, z_K\}$ denotes the hidden topics, P(d) denotes the probability of document d, P(z|d) denotes the probability of a hidden topic z given a document d, and P(v|z) denotes the probability of a word v given a hidden topic z. The objective function of the PLSA is to find the maximum likelihood (L):

$$L = \prod_{d \in D} \prod_{v \in V} P(d, v)^{n(d,v)} \tag{2}$$

where $d \in D = \{d_1, d_2, \ldots, d_M\}$, and $v \in V = \{v_1, v_2, \ldots, v_N\}$.

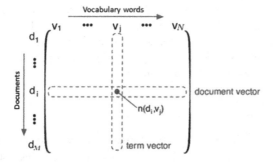

Fig. 1. Term-document matrix, modified from [14]

Maximizing the natural logarithm of the likelihood function can be expressed as follows:

$$l = \text{LogL} = \sum_{d \in D} \sum_{v \in V} n(d, v) \log P(d, v)$$
$$= \sum_{d \in D} \sum_{v \in V} n(d, v) \log P(d) \sum_{z \in Z} P(v|z)P(z|d) \tag{3}$$

The expectation-maximization (EM) algorithm is used to learn P(z|d) and P(v|z) in the PLSA model [12].

E-Step:

$$P(z|d,v) = \frac{P(v|z)P(z|d)}{\sum_{z'} P(v|z')P(z'|d)} \tag{4}$$

M-Step:

$$P(v|z) = \frac{\sum_d n(d, v)P(z|d, v)}{\sum_{v'} \sum_d n(d, v')P(z|d, v')} \tag{5}$$

$$P(z|d) = \frac{\sum_v n(d, v)P(z|d, v)}{n(d)} \tag{6}$$

The EM algorithm is reiterated between E-step and M-step repeatedly until a stop condition is met. Note that the EM algorithm would compute the training P(v|z) and yet never update it during inference. In other words, only Eqs. (4) and (6) will be iterated during inference.

3   Methodology

As mentioned earlier, our goal is to develop movie genre classifiers using the concept of PLSA to combine text, audio, and visual features. However, the standard PLSA is usually used to explain the relationship between documents and words, and cannot be applied to either audio data type or visual data type directly. Therefore, Sect. 3.1 first illustrates how the standard PLSA is implemented to unimodal PLSAs, which use either text, audio, or visual features only. Next, double-model PLSAs are extended to combine two types of information in Sect. 3.2. Finally triple-model PLSAs are developed to handle three different aspects of information.

3.1   Unimodal PLSAs

A unimodal PLSA uses one data type of information only and can be depicted in Fig. 2. There are three different unimodal PLSAs, namely, tag_based, audio_based, and visual_based. We describe each unimodal PLSA in detail as follows.

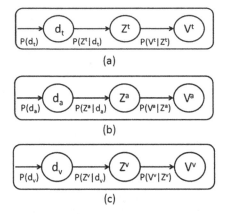

Fig. 2. Three different unimodal PLSAs. (a) Tag_based, (b) audio_based, and (c) visual_based

Tag_based PLSA. The first unimodal we implemented is the tag_based PLSA. We have downloaded 140 movie previews from IMDB [10], and have also collected their corresponding social tags from [11]. The number of collected social tags was 218. Thus, the dimensions of the term-document matrix (see Fig. 1) is 140×218 for the tag_based PLSA. Figure 2(a) illustrates the unimodal of the tag_based PLSA, d_t represents documents of movie previews, Z^t represents the hidden topics, and V^t represents the vocabulary of social tags. Superscript or subscript "t" denotes the symbol is related to the text data.

Audio_based PLSA. We constructed a term-document matrix (see Fig. 1) for the audio-based PLSA. We have chosen 10,000 frames at random for each preview, and extracted the audio features of volume, pitch, ZCR (Zero Crossing Rate), and 13 MFCC (Mel-frequency cepstral coefficient) dimensions for each frame. For more detail of visual feature extractions, simply refer to any audio textbooks or see [6, 15, 16]. Thus, the audio features for a document of preview, d_a, may be expressed as follows:

$$d_a = f_1, f_2, f_3, \ldots, f_{10000} \tag{7}$$

where f_i represents the i-th frame, and each frame contains the audio features of pitch, volume, ZCR, and 13D of MFCC. A frame may be considered as an audio word and therefore a document of preview represented by 10,000 audio words. Thus, the total number of audio words is $14 * 10^5$ for the collected set of 140 previews. To reduce the great number of audio words, we have applied a data clustering algorithm to group similar audio words. Thus, all the audio words in a same group are considered the same. The data clustering algorithm we chose is the self-organizing map (SOM). A 2D SOM with a size of 30×30 was used in our experiment. In other words, there are altogether 900 (30×30) groups. Each frame (audio word) was then mapped onto one of 900 groups according to SOM. Because the number of audio words is reduced to 900, the size of the term-document matrix (see Fig. 1) would be 140×900 for the audio_based PLSA. The unimodal for the audio signal is shown in Fig. 2(b). In this unimodal PLSA, d_a represents the movie preview documents, Z^a represents the hidden topics, and V^a represents the vocabulary of the clustering audio words. Superscript or subscript "a" denotes that the symbol is related to the audio signal.

Visual_based PLSA. A video can be considered a collection of frames. The visual features of each frame we extracted are motion, colors (RGB), and lighting. Each frame is considered a visual word. Because lengths of the collected previews are quite different, the number of frames for different previews will not be the same. This is very similar to documents, in which they also have different lengths. However, we can consider that all the words of the various documents come from the same collection of vocabulary words. To obtain the vocabulary of the visual words, a clustering algorithm is used. Similar to the procedure of audio_based PLSA, a 2D SOM with a size of 30×30 was used to find the clustering visual words. Thus, the size of the term-document matrix (see Fig. 1) is also 140×900 for the visual_based PLSA. The unimodal for the video signal is shown in Fig. 2(c). In this unimodal PLSA, d_v represents documents of movie previews, Z^v represents the hidden topics, and V^v represents the vocabulary of the clustering visual words. Either the superscript or the subscript "v" denotes that the symbol is related to the video data.

3.2   Double-Model PLSA

A double-model PLSA is used to incorporate any two different data types. For example, Fig. 3 depicts a double-model PLSA whose features are drawn from two types of information: text and audio. Firstly, we perform a tag_based PLSA and an audio_based PLSA separately and then combine the resulting $P(z^t|d_t)$ and $P(z^a|d_a)$ to generate a new joint vocabulary V^{t+a}. Both of the values, $P(z^t|d_t)$ and $P(z^a|d_a)$, represent the probability of a hidden topic z given a document d. The only difference between them are their data types, tags and audio data. Therefore, we think this is a reasonable approach to fuse two different types of information. We perform the standard (unimodal) PLSA again and use the joint vocabulary V^{t+a} as a set of words and still use the set of the collected previews as documents. Similarly, we have a double-model for incorporating text and visual data types, and a double-model for incorporating audio and visual data types.

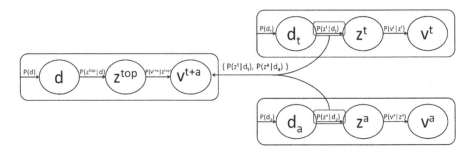

Fig. 3. A double-model PLSA illustrated by combining tag and audio data types

3.3 Triple-Model PLSA

A triple-model PLSA is used to incorporate three different aspects of information: text, audio, and visual features. We propose two categories of triple-model PLSAs. The first category of the triple-model PLSA is shown in Fig. 4(a). This is what we call tag_audio_visual PLSA. Firstly, this triple-model PLSA performs tag_based, audio-based, and visual-based PLSA separately and then combines the resulting $P(z^t|d_t)$, $P(z^a|d_a)$, and $P(z^v|d_v)$ to generate a new joint vocabulary V^{t+a+v}. Lastly, the standard (unimodal) PLSA is performed using both the joint vocabulary V^{t+a+v} as a set of words and the set of the collected previews as documents.

The second category contains three variants of triple-model PLSA. The model shown in Fig. 4(b) is what we call tag_merged_last triple-model. This model performs a double-model PLSA for combining audio and visual features, and then merges the tag features. Thus, the first double-model PLSA combines audio and visual data types, then the second double-model PLSA combines the data type of social tag and the resultant audio_visual features from the first double-model. The other two triple-model PLSAs in the second category are audio_merged_last and visual_merged_last.

4   Experimental Results

We have collected 140 movie previews from [10] and their corresponding social tags from [11]. The collected previews are divided into four movie genres: action, biography, comedy, and horror. We consider that a movie preview constitutes three different modalities: text, audio, and visual. In the experiments, we perform three unimodal PLSAs, three double-model PLSAs, and four triple-model PLSAs. The PLSA variants are used to extract topic features from one, two, or three modalities, and the naive Bayes classifier is used for classification. Figure 5 shows the classifying accuracies of movie genres using different models of PLSA.

In Fig. 5, the first three experiments denote the performance of three unimodal PLSAs: tag_based, audio_based, and visual_based. Each unimodal PLSA uses a single data type only. It is very obvious that the result of tag_based is superior to that of both audio_based and visual_based.

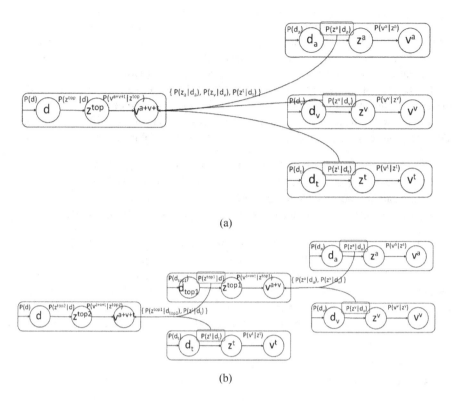

Fig. 4. Two categories of triple-model PLSAs. (a) tag_audio_visual PLSA: the only model in the first category, and (b) tag_merged_last triple-model: one of three variants model in the second category; the other two variants are audio_merged_last and visual_merged_last

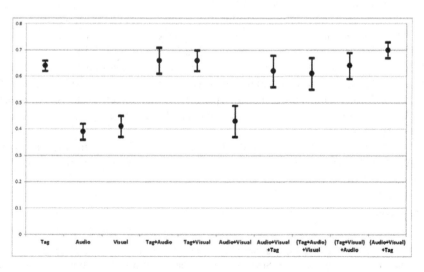

Fig. 5. Comparisons of various PLSA models

The next three experiments are for the double-model PLSAs that combine two different data types. There are three cases: combining tag and audio data types (Tag + Audio), combining tag and visual data types (Tag + Video), and combining audio and visual data types (Audio + Visual). The accuracies of case 1 and case 2 are both 66 %, which is much higher than that of case 3 (43 %). Moreover, we have observed that these two cases of double-model PLSAs contains the data type of social tags.

The last four experiments are for triple-model PLSAs from two different categories. There are four cases altogether and they are described as follows.

Case 1: tag_audio_visual PLSA (see Fig. 4(a)), (Tag + Audio + Video).
Case 2: visual_merged_last model, ((Tag + Audio) + visual).
Case 3: audio_merged_last model, ((Tag + visual) + Audio).
Case 4: tag_merged_last model (see Fig. 4(b)), ((Audio + visual) + Tag).

It can be seen that case 4 outperforms the other cases of triple-model PLSAs.

For the performance of all the variants of PLSAs, we also observed that the performance was poor if the PLSA model not containing the tag data type. More precisely, the results of two unimodal PLSAs (audio_based, and visual_based) and the double-model for audio_visual data types are unsatisfactory. This indicates that social tags play an important role for classifying movies genres. Finally, we found that the tag_merged_last model PLSA achieved the highest degree of accuracy.

5   Conclusions

In this paper, our aim is to combine text, audio and visual data types for classifying movie previews into different genres (action, biography, comedy, and horror). The PLSA is used to combine different types of information and to extract hidden topics. We have introduced and compared various PLSA approaches on the set of 140 collected movie previews and their social tags. The experimental results indicate that the triple-model PLSAs perform well in movie genre classification. In particular, the tag_merged_last model PLSA that first combines audio and visual features and then merges the tag feature has achieved the highest accuracy level. This result demonstrates that the PLSA can combine different data types for information retrieval. From the results of all the variants of PLSAs, we have also found that social tags play an important role for classifying movies genres.

In the future, we will focus on the study of the feature extractions for audio and video data types and the parameter settings of triple-model PLSAs to improve the performance of the movie genre classifications.

References

1. Brezeale, D., Cook, D.J.: Using closed captions and visual features to classify movies by genre. In: Poster Session of the Seventh International Workshop on Multimedia Data Mining (2006)

2. Moncrieff, S., Venkatesh, S., Dorai, C.: Horror film genre typing and scene labeling via audio analysis. In: Multimedia and Expo, 2003. ICME'03. Proceedings. 2003 International Conference on IEEE vol. 2, pp. II–193 (2003)
3. Arijon, D.: Grammar of the Film Language. Focal Press, London (1976)
4. Nam, J., Tewfik, A.H.: Combined audio and visual streams analysis for video sequence segmentation. In: 1997 IEEE International Conference on Acoustics, Speech, and Signal Processing. Munich (1997)
5. Pfeiffer, S., Fischer, S., Effelsberg, W.: Automatic audio content analysis. In: Proceedings of the fourth ACM international conference on Multimedia, New York, NY, pp. 21–30 (1997)
6. Wang, Y., Huang, J., Liu, Z., Chen, T.: Multimedia content classification using motion and audio information. In: Proceedings of 1997 IEEE International Symposium on Circuits and Systems on IEEE, vol. 2, pp. 1488–1491 (1997)
7. Liu, Z., Wang, Y., Chen, T.: Audio feature extraction and analysis for scene segmentation and classification. J. VLSI Sig. Proc. Syst. Sig., Image Video Technol. 20(1-2), 61–79 (1998)
8. Jain, S.K., Jadon, R.S.: Movies genres classifier using neural network. In: 24th IEEE International Symposium on Computer and Information Sciences. Guzelyurt (2009)
9. Rasheed, Z., Shah, M.: Movie genre classification by exploiting audio-visual features of previews. In: 16th International Conference on Pattern Recognition, IEEE, vol. 2, pp. 1086–1089 (2002)
10. Internet Movie Database: http://www.imdb.com/
11. Douban Movies: http://movie.douban.com/
12. Hofmann, T.: Unsupervised learning by probabilistic latent semantic analysis. Mach. Learn. 42, 177–196 (2001)
13. Levy, M., Sandler, M.: Music information retrieval using social tags and audio. IEEE Trans. Multimedia 11(3), 383–395 (2009)
14. Lienhart, R., Romberg, S., Horster, E.: Multilayer PLSA for multimodal image retrieval. In: Proceedings of ACM International Conference on Image and Video Retrieval. New York, NY, USA (2009)
15. Jang, J.S.: Audio signal processing and recognition. Available at the links for on-line courses at: http://jang/books/audioSignalProcessing
16. Zhang, T., Jay Kuo, C.-C.: Audio content analysis for online audiovisual data segmentation and classification. In: IEEE Transactions on Speech and Audio Processing (2001)

One-and-a-Half-Class Multiple Classifier Systems for Secure Learning Against Evasion Attacks at Test Time

Battista Biggio[1(✉)], Igino Corona[1], Zhi-Min He[2], Patrick P.K. Chan[2], Giorgio Giacinto[1], Daniel S. Yeung[2], and Fabio Roli[1]

[1] Department of Electrical and Electronic Engineering, University of Cagliari, Piazza d'Armi, 09123 Cagliari, Italy
{battista.biggio,igino.corona,giacinto,roli}@diee.unica.it
[2] School of Computer Science and Engineering, South China University of Technology, Guangzhou, China
{zhiminhe,patrickchan,danyeung}@ieee.org

Abstract. Pattern classifiers have been widely used in adversarial settings like spam and malware detection, although they have not been originally designed to cope with intelligent attackers that manipulate data at test time to evade detection. While a number of adversary-aware learning algorithms have been proposed, they are computationally demanding and aim to counter specific kinds of adversarial data manipulation. In this work, we overcome these limitations by proposing a multiple classifier system capable of improving security against evasion attacks at test time by learning a decision function that more tightly encloses the legitimate samples in feature space, without significantly compromising accuracy in the absence of attack. Since we combine a set of one-class and two-class classifiers to this end, we name our approach one-and-a-half-class (1.5C) classification. Our proposal is general and it can be used to improve the security of any classifier against evasion attacks at test time, as shown by the reported experiments on spam and malware detection.

1 Introduction

Pattern recognition systems have been largely employed in security-sensitive settings like biometric identity recognition, intrusion and malware detection, spam filtering, web-page ranking and network protocol verification, to discriminate between legitimate and malicious samples. However, these applications are characterized by the presence of intelligent adversaries who can deliberately attack the classifier by carefully manipulating malicious data at test time to evade detection. From the learning perspective, this means that the underlying class-conditional probability distribution of the malicious data may change from training to test time, *i.e.*, it is subject to an *adversarial* drift [1–8].

Accordingly, pattern classifiers are often characterized by an unsatisfying *trade-off* between accuracy and security against *evasion at test time*, especially in high-dimensional feature spaces. While two-class classifiers may achieve high accuracy in the absence of attack, they can be evaded by malicious samples

F. Schwenker et al. (Eds.): MCS 2015, LNCS 9132, pp. 168–180, 2015.
DOI: 10.1007/978-3-319-20248-8_15

that are sufficiently different from the training data. This is due to the fact that these classifiers minimize the classification *risk* (or error) over the training data, assuming a stationary distribution. Therefore, it does not make any significant difference in terms of risk if regions of the feature space which are not densely populated by training data are classified as legitimate or malicious [9–11]. On the other hand, one-class classifiers (trained on legitimate data) have been exploited in security applications exactly to detect these outlying, anomalous attacks. However, one-class classifiers may exhibit a significantly lower accuracy in the absence of attack (in particular, in high-dimensional feature spaces), as they do not exploit any information on the (available) malicious training data [10].

Motivated by the complementarity of the aforementioned approaches, in this work we propose a Multiple Classifier System (MCS) that combines two-class and one-class classifiers to achieve a better trade-off between accuracy and security against evasion. For this reason, we name it one-and-a-half-class (1.5C) MCS (Sect. 3). Our MCS is able to learn a more *secure* decision function by providing a tighter enclosing of the legitimate data in feature space, while also exploiting information from the available malicious data to retain high accuracy in the absence of attack. Compared to other secure learning techniques [12–14], we do not make any specific assumption on the kind of adversarial drift that may occur at test time, but rather only *agnostically* assume that malicious data may appear everywhere in feature space at test time with a non-null probability. This also allow us to reduce the computational complexity during the training phase.

To better motivate our proposal, we provide a simplified analysis of how the classification *risk* changes in the presence of evasion attacks (Sect. 2). Then, we evaluate the security of our approach in a fair, well-principled way, exploiting a recently-proposed framework to design well-crafted evasion attacks against each given classifier, including the proposed MCS, assuming perfect and limited knowledge of the targeted system [6–8] (Sect. 4). We finally evaluate the soundness of our approach on two real-world application examples involving spam and malware detection (Sect. 5), and conclude the paper by discussing related work (Sect. 6) and future research directions (Sect. 7).

2   A Simplified Risk Analysis Under Evasion Attacks

In this section, we provide an analysis of the evasion setting under the risk minimization framework [15]. In the classical setting, it is assumed that an underlying probability distribution $p(\boldsymbol{x}, y)$ generates data samples $\boldsymbol{x} \in \mathcal{X}$ along with their class labels $y \in \mathcal{Y}$, and risk minimization amounts to finding an hypothesis $f : \mathcal{X} \mapsto \mathcal{Y}$ that minimizes the expected risk (or loss) $\ell(y, f(\boldsymbol{x}))$ over p, i.e., $f = \arg\min_{f'} R(f') = \mathbb{E}_{\boldsymbol{x}, y \sim p} \{\ell(y, f'(\boldsymbol{x}))\}$. For instance, if ℓ is the 0–1 loss, $R(f)$ corresponds to the minimum classification error, being f the optimal hypothesis that would be obtained if p were known.[1]

---

[1] Typically, the underlying process p is not known, and we are only given a finite set of samples ideally drawn from it. Then, the task of learning amounts to minimizing a trade-off between the *empirical* risk computed on such set and a regularization term (or a restricted class of functions) to avoid overfitting [15].

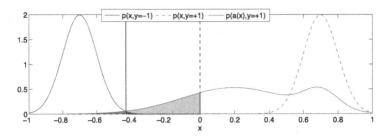

Fig. 1. Classification risk under evasion, in a one-dimensional feature space. The optimal hypothesis f learned on $p(x,y)$ classifies a sample as malicious if $x \geq 0$. Its decision boundary, achieving perfect separation between the two classes, is shown as a dashed black line. Under attack, the malicious distribution changes as $p(a(x), y = +1)$, and this causes an increase of the evasion rate given by the light and dark red areas. The decision boundary of the optimal function h trained after attack is depicted as a solid black line. It trades a much smaller evasion rate (dark red area) for a slightly higher false positive rate (blue area) (Color figure online).

In the evasion setting, the malicious data may change at test time, while the legitimate samples can be considered stationary, *i.e.*, not affected by adversarial drift. This behavior can be modeled with a function $a : \mathcal{X} \mapsto \mathcal{X}$ that represents how the attacker modifies the malicious samples drawn from p at test time. The additional risk incurred by f at test time can be thus written as:

$$R_{\text{ts}}(f) - R_{\text{tr}}(f) = \mathbb{E}_{x,y}\{\ell(y, f(a(x))) - \ell(y, f(x))\}, \tag{1}$$

where $R_{\text{tr}}(f)$ and $R_{\text{ts}}(f)$ respectively represent the risk incurred by f before and after the attack. As the legitimate data is not affected by the function $a(x)$, the above difference is not null only for the malicious class. Thus, if ℓ is the 0–1 loss, it corresponds to the increase in the evasion rate at test time (see Fig. 1).

Now, assume that $f^\star = \arg\min_{f' \in \mathcal{F}} R_{\text{ts}}(f')$ is the optimal hypothesis on the test data, including the manipulated attacks. Then, the difference of the risk incurred at test time by using f instead of $f^\star$ is:

$$R_{\text{ts}}(f) - R_{\text{ts}}(f^\star) = \mathbb{E}_{x,y}\{\ell(y, f(a(x))) - \ell(y, f^\star(a(x)))\}. \tag{2}$$

If ℓ is the 0–1 loss, this amounts to the variation of the classification error between f and $f^\star$, computed on the manipulated test data.

As shown in Fig. 1, one may thus decide to trade a slightly higher number of misclassified legitimate samples for a significantly reduced evasion rate (*i.e.*, fraction of misclassified malicious samples after attack), improving security at the expense of slightly worsening accuracy in the absence of attack. To this end, decision boundaries that more tightly enclose the legitimate class should be designed by classifying regions for which $p(x) \approx 0$ as malicious. This should allow for improving security under attack by reducing the feasible attack space (*i.e.*, regions of the feature space classified as legitimate), without significantly increasing the rate of misclassified legitimate samples. Notably, this can be a

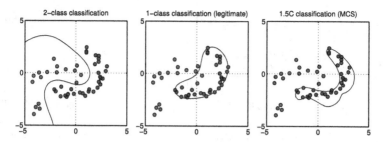

Fig. 2. 1.5C classification on two-dimensional toy data. Legitimate (malicious) training samples are shown as blue (red) points, and decision boundaries as solid black lines. *Left*: Two-class classification yields high accuracy in the absence of attack, but it can be evaded by evasion samples that are sufficiently different from the training data (*e.g.*, in the top-left corner). *Middle*: One-class classification (on the legitimate class) may worsen accuracy in the absence of attack, but improves security by enclosing the legitimate class. *Right*: 1.5C classification retains the advantages of both approaches: security is improved by enclosing the legitimate data, without significantly affecting accuracy in the absence of attack (Color figure online).

relevant problem especially in high-dimensional feature spaces, where regions that are classified as legitimate despite $p(x) \approx 0$ may be significantly wider.

3   Secure 1.5C Classification with MCSs

The analysis reported in the previous section suggests that two-class and one-class classifiers can be considered as complementary techniques, usually characterized by different challenges, especially in high-dimensional feature spaces, and a different trade-off between accuracy and security against evasion at test time. Towards *enhancing* this trade-off, we propose an MCS architecture where a two-class classifier is combined with two one-class classifiers, learned respectively on legitimate and malicious data: the two-class classifier should allow for high accuracy in the absence of attack, while the two one-class classifiers should enable the detection of evasion attacks that are (expected to be) significantly different from the training samples of either class. To combine the given classifiers in a *secure* way, *i.e.*, learning a decision function that encloses the legitimate data, we use a further one-class classifier trained on the outputs of the three base classifiers using only legitimate data. The trade-off between accuracy and security exhibited by one-class and two-class classification is exemplified in Fig. 2, along with an example of 1.5C classification overcoming the limitations of both approaches. The architecture of the proposed MCSs is depicted in Fig. 3.

4   Classifier Evasion

In this section, we consider a simplified version of the evasion attack algorithm proposed in [8], described in terms of the attack framework originally defined

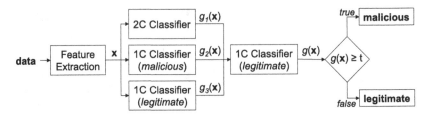

Fig. 3. 1.5C MCS: (*i*) features are extracted from raw data; (*ii*) the two-class and one-class classifiers assign scores $g_i(\boldsymbol{x}) \in \mathbb{R}$, $i = 1, 2, 3$, to the feature vector $\boldsymbol{x}$ (assuming that higher scores are given to the malicious class); (*iii*) these scores are combined by the one-class combiner, providing an aggregated score $g(\boldsymbol{x})$; (*iv*) $g(\boldsymbol{x})$ is then compared against a decision threshold t to make the final decision (Color figure online).

in [6], *i.e.*, by clearly stating the attacker's goal, knowledge of the system, and capability of manipulating the input data.

Attacker's Goal. In an evasion attack, the attacker's goal is to have malicious samples misclassified as legitimate at test time.

Attacker's Knowledge. The attacker may have different levels of knowledge about (*i*) the training data, (*ii*) the feature representation, and (*iii*) the learning model (including knowledge of the classifier's parameters after training, and feedback on its decisions on input samples). As in previous work [6,8], we consider here *perfect knowledge* (PK) and *limited knowledge* (LK) attacks. In the PK case, the attacker is assumed to know all the system details, including the trained classifier's parameters (*e.g.*, the weights assigned by a linear classifier to each feature). Although this may not be very realistic in practice, performing a security evaluation of the system under this setting allows one to assess the worst performance degradation that may be incurred by the system under attack. In the LK case, the attacker does not have access to the training data, but can collect surrogate data ideally sampled from the same distribution, and use feedback on the classifier's decisions to re-label such samples. The feature representation and the learning model (but not the trained classifier's parameters) are known. Under this setting, the attacker can learn a *surrogate* classifier $\hat{g}(\boldsymbol{x})$ on the re-labeled surrogate data, to approximate the discriminant function $g(\boldsymbol{x})$ of the targeted classifier.

Attacker's Capability. In an evasion attack, the attacker is assumed to be able to modify malicious data at test time, according to application-specific constraints, while she can neither access nor modify the classifier's training data.

Attack Strategy. Similarly to [8], and according to the previously-described attack scenario, an optimal evasion strategy can be thus formulated as:

$$\boldsymbol{x}^* = \arg\min_{\boldsymbol{x}} \quad \hat{g}(\boldsymbol{x}),$$
$$\text{s.t.} \quad d(\boldsymbol{x}, \boldsymbol{x}_a) \leq d_{\max}, \tag{3}$$

where the attacker's goal is to find a set of feasible manipulations to the initial malicious sample x_a to obtain a camouflaged sample x^* that minimizes the (surrogate) classifier's discriminant function $\hat{g}(x)$. Note that, without loss of generality, we are assuming here that the classification function is $f(x) = (g(x))_+$, where $(a)_+ = +1$ if $a \geq 0$, and -1 otherwise, and malicious samples should be assigned higher (positive) values of $g(x)$ to be correctly classified.

The feasible domain for x is defined in terms of constraints on the manipulation of the feature values of the malicious samples. For some classes of features, a simple distance metric can be adopted. For instance, if one considers the presence or absence of a given word in an email as a feature, then the ℓ_1-norm between two feature vectors amounts to counting how many words are different between the two emails. Therefore, as also done in previous work [1,2,5,6,8,16], we express this constraint as $d(x, x_a) \leq d_{\max}$, highlighting that only a maximum number of modifications $d_{\max}$ to each sample are allowed. Varying the parameter $d_{\max}$ is fundamental to properly understand system security under attack. In fact, as we will show in our experiments, more secure classifiers will exhibit a lower decrease in the detection rate of malicious samples as $d_{\max}$ increases (*i.e.*, as a larger amount of manipulations are performed to the malicious samples). Additional constraints to Problem (3) may be also considered, depending on the specific feature representation; *e.g.*, if features are real-valued and normalized in $[0,1]^d$, one may consider an additional box constraint given as $x \in [0,1]^d$.

The solution of Problem (3) clearly depends on the kind of discriminant function and on the given set of constraints. In [8], a general solution for *nonlinear*, differentiable discriminant functions was proposed, based on a simple gradient-descent algorithm. It is worth remarking that an additional component to the objective function was also considered in that work, to drive the attack point during the descent towards regions of the feature space more densely populated by legitimate samples, with the goal of increasing the probability of evading detection (especially when attacking the surrogate classifier). In this work, we disregard this component, as it is possible to substantially obtain the same effect by running the evasion attack algorithm (given as Algorithm 1) from distinct initialization points, eventually retaining the attack sample that achieves the minimum value of the objective.

4.1   Gradient Computation

The gradients required to evaluate classifier security using Algorithm 1 are given below, for the classifiers considered in this work: linear and nonlinear Support Vector Machines (SVMs), and the proposed 1.5C MCS. Note that this approach can be easily applied to any classifier, provided that its discriminant function $g(x)$ is at least subdifferentiable; otherwise, one may use a different optimization strategy to solve Problem (3), or approximate the non-differentiable $g(x)$ with a *surrogate* classifier having a differentiable discriminant function [8].

Linear Classifiers. For linear discriminant functions $g(\mathbf{x}) = \langle \mathbf{w}, \mathbf{x} \rangle + b$, with feature weights $\mathbf{w} \in \mathbb{R}^d$ and bias $b \in \mathbb{R}$, the gradient is simply $\nabla g(\mathbf{x}) = \mathbf{w}$.

Algorithm 1. Evasion Attack

Input: $\boldsymbol{x}_a$: the malicious sample; $\boldsymbol{x}^{(0)}$: the initial location of the attack sample; t: the gradient step size; ϵ: a small positive constant.
Output: $\boldsymbol{x}'$: the evasion attack sample.
1: $i \leftarrow 0$
2: **repeat**
3:    $i \leftarrow i + 1$
4:    $\boldsymbol{x}^{(i)} \leftarrow \boldsymbol{x}^{(i-1)} - t\nabla g(\boldsymbol{x}^{(i-1)})$
5:    **if** $d(\boldsymbol{x}^{(i)}, \boldsymbol{x}_a) > d_{\max}$ or other constraints are violated **then**
6:        Project $\boldsymbol{x}^{(i)}$ onto the feasible domain
7:    **end if**
8: **until** $|g(\boldsymbol{x}^{(i)}) - g(\boldsymbol{x}^{(i-1)})| < \epsilon$
9: **return** $\boldsymbol{x}' = \boldsymbol{x}^{(i)}$

Nonlinear SVMs. For kernelized (and one-class) SVMs, the discriminant function is $g(\mathbf{x}) = \sum_{i=1}^{n} \alpha_i y_i k(\mathbf{x}, \mathbf{x}_i) + b$, where $\boldsymbol{\alpha}$ and b are the SVM parameters learned during training, $k(\mathbf{x}, \mathbf{x}_i)$ is the kernel function, and $\{\boldsymbol{x}_i, y_i\}_{i=1}^{n}$ are the training samples and their labels [15]. The gradient is $\nabla g(\mathbf{x}) = \sum_i \alpha_i y_i \nabla k(\mathbf{x}, \mathbf{x}_i)$ and, thus, the feasibility of our approach depends on the kernel gradient $\nabla k(\mathbf{x}, \mathbf{x}_i)$, which is computable for many numeric kernels; e.g., for the RBF kernel $k(\mathbf{x}, \mathbf{x}_i) = \exp\left(-\gamma\|\mathbf{x} - \mathbf{x}_i\|^2\right)$, it is $\nabla k(\mathbf{x}, \mathbf{x}_i) = -2\gamma \exp\left(-\gamma\|\mathbf{x} - \mathbf{x}_i\|^2\right)(\mathbf{x} - \mathbf{x}_i)$.

1.5C MCS. In this case, the gradient will depend on the particular choice of each of the component classifiers g_1, g_2, g_3 and on the one-class combiner $g(\boldsymbol{x})$. To provide an example, we assume here that a one-class SVM with the RBF kernel is used to combine g_1, g_2, g_3. In this case, the discriminant function can be written as $g(\boldsymbol{x}) = \sum_i \alpha_i \exp\left(-\gamma\|\boldsymbol{z}(\boldsymbol{x}) - \boldsymbol{z}(\boldsymbol{x}_i)\|^2\right) + b$, where $\boldsymbol{z}(\boldsymbol{x}) = [g_1(\boldsymbol{x}), g_2(\boldsymbol{x}), g_3(\boldsymbol{x})]^\top$. The corresponding gradient is thus given as:

$$\nabla g(\boldsymbol{x}) = -2\gamma \sum_i \alpha_i e^{-\gamma\|\boldsymbol{z}(\boldsymbol{x}) - \boldsymbol{z}(\boldsymbol{x}_i)\|^2} (\boldsymbol{z}(\boldsymbol{x}) - \boldsymbol{z}(\boldsymbol{x}_i))^\top \frac{\partial \boldsymbol{z}}{\partial \boldsymbol{x}}, \qquad (4)$$

where $\frac{\partial \boldsymbol{z}}{\partial \boldsymbol{x}} = [\nabla g_1(\boldsymbol{x}), \nabla g_2(\boldsymbol{x}), \nabla g_3(\boldsymbol{x})]^\top \in \mathbb{R}^{3 \times d}$, and d is the feature set size.

Descent in Discrete Spaces. In discrete spaces, it is not possible to follow the gradient-descent direction precisely, as it may map the given sample to a set of non-admissible feature values. The gradient-descent direction can be nevertheless used as a search heuristic, as follows. Starting from the current sample $\boldsymbol{x}$, one may generate a set of candidate neighboring points by perturbing only those features of the current sample which correspond to the maximum absolute values of the gradient, one at a time, in the correct direction. Eventually, one should update the current sample to the neighbor that attained the minimum value of the objective function, and iterate until convergence.

5   Experiments

In this section, we report an experimental evaluation of our proposal on two distinct security applications, i.e., spam filtering and PDF malware detection.

Our goal is to investigate whether, and to what extent, the proposed 1.5C MCS is able to combine the advantages of two-class and one-class classification to yield a better trade-off between accuracy and security. To this end, we adopt a common experimental setup, involving the following classifiers: (*i*) a two-class SVM, with the linear kernel for the spam data, and the RBF kernel for the PDF data; (*ii-iii*) two one-class SVMs with the RBF kernel, trained respectively on legitimate and malicious samples; and (*iv*) a one-class SVM with the RBF kernel to combine the former three classifiers in our 1.5C MCS (see Sect. 3).

We simulate attackers that have perfect (PK) and limited (LK) knowledge of the targeted system (see Sect. 4). In the LK case, we assume that the attacker collects a surrogate training set whose size is only 20 % of the size of the data used to learn the targeted classifier. The reason is that, in practice, attackers may only perform a limited number of queries to the targeted classifier to recover the classification labels of such samples, and train the surrogate classifier.

We evaluate classifier performance by averaging the true positive (TP) rate, *i.e.*, the fraction of correctly-classified malicious samples, achieved when the false positive (FP) rate, *i.e.*, fraction of misclassified legitimate samples, is below 1 %, which corresponds to the so-called Area Under the ROC Curve (AUC) at 1 % FP rate. The reason is that, in these applications, false positives are more harmful than false negatives, and thus, they should be kept very low [6,18]. Our results are averaged over five repetitions, in which different training-test splits are considered. In particular, in each repetition, 50 % of the data is randomly chosen for training (TR), while the rest is used for testing (TS).

5.1   Spam Filtering

Spam filtering is one of the most popular application examples considered in adversarial learning [1,2,6,17,18]. As in previous work, we consider the widely-used bag-of-words feature representation, in which an email is described as a set of binary features, indicating the presence (1) or absence (0) of a given word. Due to the high dimensionality of the corresponding feature space, linear classifiers have been often considered [17–19] and implemented in real anti-spam filters, including *e.g.* SpamAssassin and SpamBayes.[2] Therefore, in this investigation we implement the two-class classifier as a linear SVM. Popular attacks in spam filtering include modifications to the email's content. Attackers may obfuscate *bad* words (which typically appear in spam but not in legitimate emails) through misspelling (*e.g.*, changing "cheap" as "che4p"), and add *good* words (which typically appear in legitimate emails but not in spam). The corresponding effect is to change features from 1 to 0, or viceversa.

Experimental Setup. We use the benchmark TREC 2007 email corpus [20], which consists of 25,220 ham and 50,199 spam emails. The first 5,000 emails are adopted in this experiment. Each email is represented by a feature vector using a tokenization method of SpamAssassin. We use a feature selection method

---

[2] http://spamassassin.apache.org/, http://spambayes.sourceforge.net/.

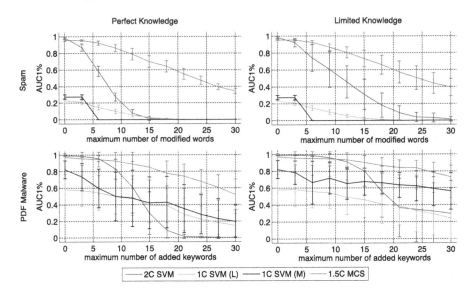

Fig. 4. Average AUC$_{1\%}$ values (with standard deviation) for spam filtering (top row) and PDF malware detection (bottom row), attained by two-class (2C) SVMs, one-class (1C) SVMs trained on legitimate (L) and malicious (M) data, and the 1.5C MCS, against PK (first column) and LK (second column) attacks. In each plot, the AUC$_{1\%}$ under attack is evaluated against an increasing maximum number of modifications to the malicious data, *i.e.*, number of modified words in each spam, and number of added keywords to each malicious PDFs.

based on information gain [21], reducing the number of features from more than 20,000 to 500, to reduce computational complexity. All the malicious samples in TS are manipulated according to PK and LK evasion attacks. In the latter case, only 500 samples have been used as the surrogate training data available to the attacker. To evaluate classifier security against an increasing maximum number of modifiable words in each spam, we set $d(\boldsymbol{x}, \boldsymbol{x}_a) = \|\boldsymbol{x} - \boldsymbol{x}_a\|_1 \le d_{\max}$ as the constraint in Problem (3). In this case, $d_{\max}$ is exactly the maximum number of modifiable words (*i.e.*, feature values) in each spam. The parameters $C, \gamma \in \{2^{-10}, 2^{-9}, \dots, 2^{10}\}$ of each SVM are optimized through a 5-fold cross-validation on TR to minimize the classification error.

Results. The average AUC$_{1\%}$ values, along with their standard deviations, are shown in Fig. 4 (top row), against an increasing maximum number of modified words in each spam. In the absence of attack, the accuracy of the 1.5C MCS and the two-class (2C) SVM is similar, and very high, while for one-class (1C) SVMs it is very low. The reason is that we are not only working at extremely low FP rates, but also in high-dimensional spaces; thus, detecting a high number of attacks for these techniques becomes extremely challenging. Under attack, *i.e.*, where the number of modified words is greater than zero, the AUC$_{1\%}$ of the 2C SVMs drops significantly, in both the PK and the LK settings. Conversely, the performance of our 1.5C MCS decreases more gracefully, retaining a higher detection rate against evasion attempts, and thus, a higher security.

5.2   PDF Malware Detection

Another well-known application investigated in adversarial learning is PDF malware detection. PDF files are characterized by a hierarchy of interconnected objects including keywords and data streams; *e.g.*, the keyword "PageLayout" characterizes an object describing the format of the corresponding page. Their flexible structure allows for embedding different kinds of content, including JavaScript and binary programs. This makes PDF files suitable vectors to spread malware infections, as malware can be easily embedded into them. In this experiment, we use the feature representation adopted in [8,22], in which each feature represents the number of occurrences of a given keyword in a PDF file. In this case, evasion attacks consist of manipulating the set of keywords present in the PDF file at test time. However, since embedded objects can be easily added to, but not removed from PDF files without corrupting their structure, as in [8,22] we assume that the number of occurrences of each keyword can be only increased. This can be implemented by adding the constraint $x_a \leq x$ (where inequality has to be fulfilled by each feature value) to Problem (3).

Experimental Setup. We use 2,000 samples from the PDF malware dataset in [8,22], represented by 114 distinct keywords (*i.e.*, features). Each feature value (*i.e.*, the number of occurrences of the corresponding keyword) is normalized in $[0, 1]$ by dividing its value by 100 (which is also set as the maximum value of the occurrences of each keyword). The SVMs' parameters are set as in the previous experiments on spam filtering. Classifier security is evaluated against an increasing maximum number of added keywords. To this end, we set $d(x, x_a) = 100\|x - x_a\|_1 \leq d_{\max}$ as the constraint in Problem (3), similarly to the experiment on spam filtering. Together with the constraint on feature addition $x_a \leq x$, this let $d_{\max}$ correspond to the maximum number of keywords that can be added to each malicious PDF file.

Results. As for the spam filtering case, results are reported in terms of $\text{AUC}_{1\%}$ values in Fig. 4 (bottom row). Similar conclusions can be drawn with respect to the results obtained for the spam filtering case, except that here, 1C SVMs perform slightly better in the absence of attack, and exhibit higher detection rates under attack than those exhibited by the 2C SVMs (for both PK and LK attacks). This means that, for a sufficiently high number of added keywords, 1C SVMs are more secure than 2C SVMs. Nevertheless, the proposed 1.5C MCS significantly outperforms the competing approaches under attack, while only performing slightly worse than the 2C SVM in the absence of attack. Worth noting, the detection rate of the 1.5C MCS decreases more gracefully in this case with the increase of the maximum number of added keywords, since the feature values of malicious samples can only be incremented.

6   Related Work

While many methods have been proposed to counter evasion attacks [1,12,14,25], they are computationally demanding and make specific assumptions on how attackers should manipulate data at test time to evade detection (see also [3–6] and references therein, for a more detailed review). Inspired by the seminal work in [1], recent work has shown that game theory can be used to model interactions between the classification system and the attacker in non-zero-sum games, yielding more secure classifiers [14]. However, satisfying conditions for the uniqueness of the equilibrium not only require the classifier's and the attacker's objectives to fulfill specific conditions, but also computationally demanding algorithms to find suitable solutions. A slightly different line of work [12,13,25] aims to learn secure classifiers (and SVMs, in particular), by minimizing a modified version of the loss function which accounts for worst-case manipulations to data at test time, including feature deletion, insertion, and rescaling. This also results in a computationally intensive training phase.

Other work has proposed countermeasures to evasion based on well-motivated heuristics, *e.g.*, that feature weights assigned by a linear classifier should be more evenly spread among the features, to require modifying more feature values to evade detection [18,23]. Linear classifiers exhibiting the aforementioned behavior have been implemented efficiently using MCSs, by averaging an ensemble of linear classifiers [18,23,26,27]. In the present work, we have shown that MCSs can also be used to learn secure, nonlinear classifiers, in an efficient manner.

7   Conclusions and Future Work

While pattern classifiers have been widely used in adversarial applications, they are often characterized by an unsatisfying trade-off between accuracy and security against evasion, especially in high-dimensional spaces: two-class classifiers may yield high accuracy but they are potentially insecure, while one-class classification may improve security at the expense of a lower accuracy in the absence of attack. Motivated by the complementarity of these approaches, we proposed a 1.5C MCS that achieves a better *trade-off* between accuracy and security. Our proposal is *general*, it is *agnostic* to the kind of adversarial drift that may occur at test time, and can be used to improve the security of *any classifier* against evasion attacks, as shown by the reported experiments on spam and malware detection. Interesting future research directions include providing a formal, well-principled characterization of the trade-off between accuracy and security, as well as its relationship with the dimensionality of the feature space, and investigating whether the proposed 1.5C MCS can also be capable to deal with the presence of *poisoning* attacks in the training data [3–7].

Acknowledgments. This work has been partly supported by the projects CRP-18293 and CRP-59872, both funded by Regione Autonoma della Sardegna, L.R. 7/2007, respectively with Bando 2009 and Bando 2012.

References

1. Dalvi, N., Domingos, P., Mausam, Sanghai, S., Verma, D.: Adversarial classification. In: 10th International Conference on Knowledge Discovery and Data Mining. ACM, pp. 99–108 (2004)
2. Lowd, D., Meek, C.: Adversarial learning. In: 11th International Conference on Knowledge Discovery and Data Mining. ACM, pp. 641–647 (2005)
3. Barreno, M., Nelson, B., Sears, R., Joseph, A.D., Tygar, J.D.: Can machine learning be secure? In: ASIA CCS. ACM, pp. 16–25 (2006)
4. Barreno, M., Nelson, B., Joseph, A., Tygar, J.: The security of machine learning. Mach. Learn. **81**, 121–148 (2010)
5. Huang, L., Joseph, A.D., Nelson, B., Rubinstein, B., Tygar, J.D.: Adversarial machine learning. In: 4th Workshop Artificial Intelligence and Security. ACM, pp. 43–57 (2011)
6. Biggio, B., Fumera, G., Roli, F.: Security evaluation of pattern classifiers under attack. IEEE Trans. Knowl. Data Eng. **26**(4), 984–996 (2014)
7. Biggio, B., Fumera, G., Roli, F.: Pattern recognition systems under attack: design issues and research challenges. Int. J. Pattern Recogn. Artif. Intell. **28**(7), 21 (2014)
8. Biggio, B., Corona, I., Maiorca, D., Nelson, B., Šrndić, N., Laskov, P., Giacinto, G., Roli, F.: Evasion attacks against machine learning at test time. In: Blockeel, H., Kersting, K., Nijssen, S., Železný, F. (eds.) ECML PKDD 2013, Part III. LNCS, vol. 8190, pp. 387–402. Springer, Heidelberg (2013)
9. Gori, M., Scarselli, F.: Are multilayer perceptrons adequate for pattern recognition and verification? IEEE TPAMI **20**(11), 1121–1132 (1998)
10. Tax, D.M.J.: One-class classification. Ph.D. thesis (2001)
11. Biggio, B., Fumera, G., Roli, F.: Design of robust classifiers for adversarial environments. In: IEEE International Conference on Systems, Man, and Cybernetics, pp. 977–982 (2011)
12. Globerson, A., Roweis, S.: Nightmare at test time: robust learning by feature deletion. In: 23rd ICML, vol. 148, pp. 353–360. ACM (2006)
13. Teo, C.H., Globerson, A., Roweis, S., Smola, A.: Convex learning with invariances. In: Platt, J., et al., eds.: NIPS 20, pp. 1489–1496. MIT Press (2008)
14. Brückner, M., Kanzow, C., Scheffer, T.: Static prediction games for adversarial learning problems. J. Mach. Learn. Res. **13**, 2617–2654 (2012)
15. Vapnik, V.N.: The nature of statistical learning theory. Springer, New York (1995)
16. Nelson, B., Rubinstein, B.I., Huang, L., Joseph, A.D., Lee, S.J., Rao, S., Tygar, J.D.: Query strategies for evading convex-inducing classifiers. J. Mach. Learn. Res. **13**, 1293–1332 (2012)
17. Nelson, B. et al.: Exploiting machine learning to subvert your spam filter. In: Large-scale Exploits and Emergent Threats, USENIX, pp. 1–9 (2008)
18. Biggio, B., Fumera, G., Roli, F.: Multiple classifier systems for robust classifier design in adversarial environments. Int. J. Mach. Learn. Cyb. **1**(1), 27–41 (2010)
19. Jorgensen, Z., Zhou, Y., Inge, M.: A multiple instance learning strategy for combating good word attacks on spam filters. J. Mach. Learn. Res. **9**, 1115–1146 (2008)
20. Cormack, G.V.: Trec 2007 spam track overview. In: Voorhees, E.M., Buckland, L.P., eds.: TREC, pp. 500–274. Volume Special Publication, NIST (2007)
21. Sebastiani, F.: Machine learning in automated text categorization. ACM Comput. Surv. **34**, 1–47 (2002)
22. Maiorca, D., Corona, I., Giacinto, G.: Looking at the bag is not enough to find the bomb: an evasion of structural methods for malicious PDF files detection. In: 8th ASIA CCS, pp. 119–130. ACM (2013)

23. Kolcz, A., Teo, C.H.: Feature weighting for improved classifier robustness. In: 6th Conference on Email and Anti-spam (2009)
24. Sutton, C., Sindelar, M., McCallum, A.: Feature bagging: preventing weight under-training in structured discriminative learning. Technical report, IR-402, University of Massachusetts (2005)
25. Zhou, Y., Kantarcioglu, M., Thuraisingham, B., Xi, B.: Adversarial support vector machine learning. In: 18th International Conference on Knowledge Discovery and Data Mining, pp. 1059–1067. ACM (2012)
26. Biggio, B., Fumera, G., Roli, F.: Multiple classifier systems for adversarial classification tasks. In: Benediktsson, J.A., Kittler, J., Roli, F. (eds.) MCS 2009. LNCS, vol. 5519, pp. 132–141. Springer, Heidelberg (2009)
27. Biggio, B., Fumera, G., Roli, F.: Multiple classifier systems under attack. In: El Gayar, N., Kittler, J., Roli, F. (eds.) MCS 2010. LNCS, vol. 5997, pp. 74–83. Springer, Heidelberg (2010)

An Experimental Study on Combining Binarization Techniques and Ensemble Methods of Decision Trees

Juan J. Rodríguez[✉], José F. Díez-Pastor, Álvar Arnaiz-González, and César García-Osorio

University of Burgos, Burgos, Spain
jjrodriguez@ubu.es

Abstract. Binarization techniques deal with multiclass classification problem combining several binary classifiers. They were originally introduced for dealing with multiclass problems with methods that were only able to deal with two classes (e.g., SVM). Nevertheless, it has been shown that they can also be useful with classification methods able to deal directly with multiclass problems (e.g., decision trees), because they can improve the results. This work studies if this improvement is also possible when using ensembles of decision trees (e.g., Random Forest, Boosting) over 67 multiclass datasets. It was found that some combinations of a binarization technique and an ensemble method improve the results of the ensemble method without binarization.

Keywords: Multiclass · Binarization · One vs. one · one vs. all · Ensembles · Bagging · Boosting · Random Forest

1 Introduction

Some methods for constructing classifiers, such as support vector machines, are intrinsically for binary problems. Others, such as decision rules, were originally conceived for binary classes, but were latter easily extended to the multiclass case. One obvious way to tackle a multiclass data set using a method for binary data sets is to decompose the problem in several subproblems and combine in some ways the predictions given by the binary classifiers. These binarization techniques could be considered as multiclassifiers or ensemble methods because the final classification is obtained combining the predictions of several classifiers, although it can be argued that they are not ensembles in a strict sense [3,29]. A framework for unifying binarization and ensembles is presented in [3].

Although binarization techniques were designed for dealing with one limitation, it has been observed that it can be sensible to use them with classifiers without this limitation, because the resulting classifiers can improve the results of the classifier. This work studies if this is also possible when the classifiers that are used with binarization are ensembles of decision trees, such as Random Forest [5] or Boosting [14]. Decision trees are usually used as base classifiers

© Springer International Publishing Switzerland 2015
F. Schwenker et al. (Eds.): MCS 2015, LNCS 9132, pp. 181–193, 2015.
DOI: 10.1007/978-3-319-20248-8_16

within ensemble methods because they are fast and unstable, these are desirable in ensembles.

The two most common approaches for binarization are "one vs. all" and "one vs. one" [17]. In the first case, sometimes named more properly "one vs. the rest", there are as many classifiers as classes, each classifier predicts if the instance is of the corresponding class or of one of the other classes. In one vs. one, also named round-robin or pairwise classification [16] there are as many classifiers as pairs of classes. Once that the binary classifiers have been constructed, there is the issue of how to combine their predictions. In one vs. one several of the classifiers can predict the corresponding class, or all the classifiers could predict that the instance is in the rest of classes. In one vs. one, many classifiers will be wrong, because they will predict one of two classes for instances of any class. There are several approaches for combining this predictions [17], usually based on the probabilities assigned by the binary classifiers to each class.

When combining binarization techniques and ensemble method, one issue is in what order the methods are applied. For instance, with decision trees we can have Bagging of trees and one vs. one of trees. But if the two are used, there are two options: "Bagging of one vs. one of trees" and "one vs. one of Bagging and trees". For some ensemble methods, such as Bagging or Random Subspaces, it can seem that this order will not be important, but for others such as Boosting the order can have more importance. Hence this paper considers both orderings.

The rest of the paper is organised as follows. Section 2 presents the experimental setup. The results are analysed in Sect. 3. Finally, Sect. 4 presents the conclusions and some open research lines.

2   Experimental Setup

2.1   Data Sets

Table 1 shows the 67 data sets used in the experiments. They are all the multiclass data sets[1] used in [12]. Many of them are from the UCI Machine Learning Repository [2]. The number of examples ranges in [24, 58000], the number of features in [3, 262] and the number of classes in [3, 100]. The table also shows the percentage of examples of the classes with more and less examples. For the majority class, the percentages range in [1, 92.58], for the minority in [0.02, 33.33].

2.2   Settings

Table 2 shows the abbreviations using in the methods configurations. Two binarization techniques were used: one vs. all and one vs. one. They were used with decision trees but also with ensembles of decision trees. The used ensemble methods were Bagging [4], Random Subspaces [21], Random Forest [5], AdaBoost.M1 [13], MultiBoost [27], LogitBoost [15] and Rotation Forest [24]. Table 2 shows the used abbreviations.

---

[1] Available from http://persoal.citius.usc.es/manuel.fernandez.delgado/papers/jmlr/.

Table 1. Characteristics of the data sets.

Data set	Examples	Features	Classes	% Majority	% Minority
Abalone	4177	8	3	34.64	31.67
Annealing	898	31	5	76.17	0.89
Arrhythmia	452	262	13	54.20	0.44
Audiology-std	196	59	18	23.98	1.02
Balance-scale	625	4	3	46.08	7.84
Breast-tissue	106	9	6	20.75	13.21
Car	1728	6	4	70.02	3.76
Cardiotocography-10classes	2126	21	10	27.23	2.49
Cardiotocography-3classes	2126	21	3	77.85	8.28
Chess-krvk	28056	6	18	16.23	0.10
Conn-bench-vowel-deterding	990	11	11	9.09	9.09
Contrac	1473	9	3	42.70	22.61
Dermatology	366	34	6	30.60	5.46
Ecoli	336	7	8	42.56	0.60
Energy-y1	768	8	3	46.88	17.84
Energy-y2	768	8	3	49.87	24.61
Flags	194	28	8	30.93	2.06
Glass	214	9	6	35.51	4.21
Hayes-roth	160	3	3	40.63	19.38
Heart-cleveland	303	13	5	54.13	4.29
Heart-switzerland	123	12	5	39.02	4.07
Heart-va	200	12	5	28.00	5.00
Image-segmentation	2310	18	7	14.29	14.29
Iris	150	4	3	33.33	33.33
Led-display	1000	7	10	11.10	8.40
Lenses	24	4	3	62.50	16.67
Letter	20000	16	26	4.07	3.67
Libras	360	90	15	6.67	6.67
Low-res-spect	531	100	9	51.98	0.38
Lung-cancer	32	56	3	40.63	28.13
Lymphography	148	18	4	54.73	1.35
Molec-biol-splice	3190	60	3	51.88	24.04
Nursery	12960	8	5	33.33	0.02
Oocytes-merluccius-2f	1022	25	3	68.69	5.97
Oocytes-trisopterus-5b	912	32	3	57.57	1.54
Optical	5620	62	10	10.18	9.86

(Continued)

Table 1. *(Continued)*

Data set	Examples	Features	Classes	% Majority	% Minority
Page-blocks	5473	10	5	89.77	0.51
Pendigits	10992	16	10	10.41	9.60
Pittsburg-bridges-MATERIAL	106	7	3	74.53	10.38
Pittsburg-bridges-REL-L	103	7	3	51.46	14.56
Pittsburg-bridges-SPAN	92	7	3	52.17	23.91
Pittsburg-bridges-TYPE	105	7	6	41.90	9.52
Plant-margin	1600	64	100	1.00	1.00
Plant-shape	1600	64	100	1.00	1.00
Plant-texture	1599	64	100	1.00	0.94
Post-operative	90	8	3	71.11	2.22
Primary-tumor	330	17	15	25.45	1.82
Seeds	210	7	3	33.33	33.33
Semeion	1593	256	10	10.17	9.73
Soybean	683	35	18	13.47	1.17
Statlog-image	2310	18	7	14.29	14.29
Statlog-landsat	6435	36	6	23.82	9.73
Statlog-shuttle	58000	9	7	78.60	0.02
Statlog-vehicle	846	18	4	25.77	23.52
Steel-plates	1941	27	7	34.67	2.83
Synthetic-control	600	60	6	16.67	16.67
Teaching	151	5	3	34.44	32.45
Thyroid	7200	21	3	92.58	2.31
Vertebral-column-3clases	310	6	3	48.39	19.35
Wall-following	5456	24	4	40.41	6.01
Waveform	5000	21	3	33.92	32.94
Waveform-noise	5000	40	3	33.84	33.06
Wine	178	13	3	39.89	26.97
Wine-quality-red	1599	11	6	42.59	0.63
Wine-quality-white	4898	11	7	44.88	0.10
Yeast	1484	8	10	31.20	0.34
Zoo	101	16	7	40.59	3.96

The ensemble methods can be used without binarization, this is denoted by E. In the combination, the binarizer can be the base classifier of the ensemble and these cases are denoted by E-OVA and E-OVO. The configurations where the ensemble is the base classifier of the binarizers are denoted by OVA-E and OVO-E.

Table 2. Abbreviations used in the methods configurations.

Abbrev	Method
OVA	One vs. all
OVO	One vs. one
OV?	One of the binarization techniques: OVO or OVA
Tree	A singe decision tree
P	Trees with pruning
U	Trees without pruning
Bagg	Bagging
RndSub	Random Subspaces
RndFor	Random Forest
AdaBo	AdaBoost.M1
MulBo	MultiBoost
LogBo	LogitBoost
RW	A boosting method with *reweighting*
RS	A boosting method with *resampling*
RotFor	Rotation Forest
E	One of the ensemble methods (Bagg, RndSub...)

The experiments were performed in Weka [20], version 3.7.12. The combination of the predictions of the binary classifiers in OVA and OVO is done adding the support assigned to each class by each classifier.

Ensemble size was 100. As base classifiers J48 trees, a re-implementation of C4.5 [22], were used. There is one exception, for LogitBoost the method REPTree [9] was used. The reason is that in LogitBoost the base models are regressors, not classifiers. Trees were used with pruning or without pruning.

For the boosting variants (i.e., AdaBo, MulBo and LogBo), the experiments were done with *reweighting* (the default option in Weka) and *resampling* [14]. In the first case, the base method receives weighted training instances, in the second case the weighted instances are used to draw a sample of unweighted instances and the base classifier is constructed with the sample.

For MultiBoost, the sub-committee size was set to 10. The rest of parameters of all the used methods had the default values in Weka.

5×2-fold cross validation was used [7]. As performance measures, accuracy and kappa [17] were used; kappa is less sensitive to the class distributions. For comparing the methods with several data sets, average ranks [6] were computed. For each data set, all the compared method are sorted from best to worst, the best method is assigned a rank of 1, the second method a rank of 2, and so on.

In the case of ties, the methods are given average ranks. For instance, if four methods have the best results, they are assigned a rank of 2.5. The ranks of the different data sets are averaged for each method.

3   Results

Table 3 shows, for the considered combinations of binarization and ensemble techniques the average accuracies across all the data sets. These averages only provide a rough idea because the results of different data sets can be not commensurable and averages can be too sensitive to outliers [6,27]. Table 4 shows the averages when using kappa as the performance measure. According to these averages, the best rows (in the rotated tables) are for the ensemble methods Rnd-For and RotFor. In general, the best column is E, that is, not using binarization techniques.

Table 3. Average accuracies across all the data sets for the considered combinations of binarization and ensemble techniques. Higher values have darker backgrounds. Empty cells indicate that the corresponding combination was not considered, for reasons explained in the text.

Method	E	OVA-E	OVO-E	E-OVA	E-OVO
Tree-P	74.578	73.239	75.180		
Tree-U	73.981	73.265	74.760		
Bagg-P	79.042	79.086	77.644	79.092	77.198
Bagg-U	79.019	79.161	77.646	79.168	77.164
RndSub-P	77.920	76.650	76.412	76.595	76.515
RndSub-U	78.197	77.506	76.776	77.527	76.855
RndFor	80.029	80.157	79.500	80.076	79.074
AdaBo-RW-P	79.097	76.657	77.439	78.680	78.087
AdaBo-RW-U	79.248	76.368	77.486	78.648	77.868
AdaBo-RS-P	79.546	77.525	77.856	78.680	78.087
AdaBo-RS-U	79.431	77.410	77.835	78.648	77.868
MulBo-RW-P	79.512	76.435	77.624	78.814	78.220
MulBo-RW-U	79.577	76.441	77.735	78.899	78.108
MulBo-RS-P	79.732	77.384	77.890	78.814	78.220
MulBo-RS-U	79.647	77.383	77.904	78.899	78.108
LogBo-RW-P	76.077	75.797	73.581		
LogBo-RW-U	78.089	77.080	76.885		
LogBo-RS-P	79.211	79.524	78.622		
LogBo-RS-U	79.169	79.268	78.733		
RotFor-P	81.324	80.960		80.994	80.108
RotFor-U	81.245	81.152		80.982	80.262

Table 4. Average values for kappa across all the data sets for the considered combinations of binarization and ensemble techniques.

	E	OVA-E	OVO-E	E-OVA	E-OVO
Tree-P	0.6337	0.6145	0.6363		
Tree-U	0.6289	0.6183	0.6341		
Bagg-P	0.6882	0.6897	0.6686	0.6898	0.6636
Bagg-U	0.6890	0.6921	0.6697	0.6923	0.6648
RndSub-P	0.6556	0.6339	0.6340	0.6337	0.6374
RndSub-U	0.6636	0.6514	0.6436	0.6524	0.6465
RndFor	0.7000	0.7022	0.6922	0.7007	0.6857
AdaBo-RW-P	0.6931	0.6678	0.6719	0.6875	0.6798
AdaBo-RW-U	0.6955	0.6638	0.6733	0.6861	0.6768
AdaBo-RS-P	0.6989	0.6765	0.6758	0.6875	0.6798
AdaBo-RS-U	0.6968	0.6750	0.6759	0.6861	0.6768
MulBo-RW-P	0.6985	0.6657	0.6747	0.6889	0.6813
MulBo-RW-U	0.6998	0.6663	0.6768	0.6892	0.6794
MulBo-RS-P	0.7007	0.6751	0.6767	0.6889	0.6813
MulBo-RS-U	0.6996	0.6753	0.6767	0.6892	0.6794
LogBo-RW-P	0.6466	0.6414	0.6080		
LogBo-RW-U	0.6810	0.6692	0.6641		
LogBo-RS-P	0.6955	0.6990	0.6869		
LogBo-RS-U	0.6941	0.6961	0.6876		
RotFor-P	0.7169	0.7097		0.7107	0.6984
RotFor-U	0.7162	0.7142		0.7120	0.7031

Table 5 shows for how many data sets a combination of binarization with ensemble has better, equal or worse accuracy than the corresponding ensemble without binarization. Table 6 shows this information for kappa. In general, the advantage is for the ensemble without binarization, but there are some cases where the results clearly favour some combinations.

Table 7 shows, for each ensemble method, the average ranks from accuracy of the configurations that include that ensemble method (the ensemble without binarization and the combination with up to four binarization techniques). Table 8 shows the average ranks according to kappa. The best ranks are usually obtained without binarization techniques (E) or with OVA.

Now, the results for each ensemble method are considered. Although single decision trees are not ensembles, but their combination with binarization techniques is also included in the comparison. The configurations E-OV? are not possible with decision trees. According to the averages across all the data sets, the number of wins and losses and the average ranks the configurations from best to worst are OVO-E, E and OVA-E. The advantages of OVO-E over E are greater for accuracy than for kappa.

Table 5. Comparison of the combination of a binarization and ensemble technique with the corresponding ensemble without binarization, according to the accuracy, in terms of the number of wins (W), ties (T) and losses (L). Each cell shows W/T/L. The background color for cells is obtained from W-L.

	OVA-E	OVO-E	E-OVA	E-OVO
Tree-P	14/2/51	40/2/25		
Tree-U	25/2/40	38/1/28		
Bagg-P	52/0/15	17/0/50	52/0/15	19/0/48
Bagg-U	51/2/14	19/1/47	51/2/14	21/1/45
RndSub-P	22/3/42	11/2/54	24/3/40	17/1/49
RndSub-U	31/2/34	16/1/50	33/0/34	16/0/51
RndFor	37/0/30	18/2/47	35/2/30	17/0/50
AdaBo-RW-P	22/1/44	24/0/43	27/0/40	26/1/40
AdaBo-RW-U	18/2/47	22/0/45	26/0/41	21/1/45
AdaBo-RS-P	26/0/41	24/0/43	26/1/40	15/1/51
AdaBo-RS-U	25/0/42	25/3/39	20/0/47	18/2/47
MulBo-RW-P	18/1/48	16/1/50	24/2/41	20/1/46
MulBo-RW-U	16/3/48	18/1/48	26/0/41	23/1/43
MulBo-RS-P	21/2/44	23/1/43	22/2/43	15/1/51
MulBo-RS-U	23/1/43	26/1/40	23/0/44	13/2/52
LogBo-RW-P	35/3/29	7/0/60		
LogBo-RW-U	21/1/45	10/2/55		
LogBo-RS-P	47/0/20	22/1/44		
LogBo-RS-U	43/2/22	20/2/45		
RotFor-P	31/2/34		30/5/32	11/1/55
RotFor-U	35/2/30		31/2/34	20/0/47

For Bagging the best configurations is OVA-E followed by E-OVA. The configurations with OVO are worse than not using binarization techniques. Interestingly, the situation for Bagging is the opposite than the situation for single decision trees: OVO is the best for decision trees but the worst for Bagging, OVA is the best for Bagging but the worst for single decision trees. This means that the knowledge about the performance of binarization techniques obtained from combining single classifiers is not valid when using ensembles.

For Random Subspaces (RndSub) the best results are obtained without binarization (E). Configurations with OVA have better results than configurations with OVO.

For Random Forest the best configurations are OVA-E and E-OVA. Configurations with OVO are worse than not using binarization.

For AdaBoost (AdaBo) the results are generally better without binarization techniques. From the combinations with binarization techniques, the most favourable one is E-OVA. In this case the order in the combination is important.

Table 6. Comparison of the combination of a binarization and ensemble technique with the corresponding ensemble without binarization, according to kappa, in terms of the number of wins (W), ties (T) and losses (L). Each cell shows W/T/L. The background color for cells is obtained from W-L.

	OVA-E	OVO-E	E-OVA	E-OVO
Tree-P	15/1/51	35/1/31		
Tree-U	22/0/45	36/1/30		
Bagg-P	48/0/19	17/1/49	46/0/21	19/0/48
Bagg-U	49/1/17	16/1/50	47/1/19	20/0/47
RndSub-P	21/2/44	12/1/54	22/1/44	18/0/49
RndSub-U	27/0/40	16/0/51	29/0/38	16/0/51
RndFor	35/0/32	18/0/49	37/1/29	17/0/50
AdaBo-RW-P	22/1/44	22/0/45	27/0/40	24/0/43
AdaBo-RW-U	18/0/49	18/0/49	25/0/42	20/0/47
AdaBo-RS-P	24/0/43	22/0/45	24/1/42	16/0/51
AdaBo-RS-U	25/0/42	26/0/41	16/0/51	18/1/48
MulBo-RW-P	20/0/47	15/1/51	22/1/44	21/0/46
MulBo-RW-U	18/1/48	16/0/51	22/0/45	22/1/44
MulBo-RS-P	22/0/45	26/0/41	22/0/45	16/0/51
MulBo-RS-U	22/0/45	26/0/41	20/0/47	11/0/56
LogBo-RW-P	32/2/33	6/0/61		
LogBo-RW-U	23/1/43	10/0/57		
LogBo-RS-P	47/0/20	24/0/43		
LogBo-RS-U	45/1/21	22/1/44		
RotFor-P	30/0/37		30/1/36	12/0/55
RotFor-U	34/0/33		30/1/36	20/0/47

For instance, for AdaBo-RW the configuration with the worst ranks is OVA-E, while E-OVA was the best.

The behaviour of MultiBoost (MulBo) is similar to AdaBoost. The configurations with best results do not use binarization (E). The best configuration with binarization is E-OVA.

For LogitBoost (LogBo) there are not configurations with E-OV?. The reason is that the base method in LogitBoost is not a classifier, but a regression method. In each iteration, for each class, a regression model is constructed. In fact, LogBo could be considered as a type of OVA, because for each iteration there are as many models as classes. The two considered versions, LogBo-RW and LogBo-RS have different behaviours. In the former, the best option is not to use binarization E, in the latter the best option is OVA-E. Comparing the two versions, the best one is LogBo-RS and in this case the binarization technique is beneficial.

For Rotation Forest the configuration OVO-E is not included. The reason is that the memory requirements were excessive for the problems with 100 classes.

Table 7. Average ranks for accuracy.

	E	OVA-E	OVO-E	E-OVA	E-OVO
Tree-P	1.836	2.500	1.664		
Tree-U	1.963	2.284	1.754		
Bagg-P	3.090	1.858	4.045	1.993	4.015
Bagg-U	3.164	1.799	4.067	2.007	3.963
RndSub-P	2.172	2.687	3.791	2.731	3.619
RndSub-U	2.455	2.493	3.754	2.530	3.769
RndFor	2.627	2.343	3.746	2.373	3.910
AdaBo-RW-P	2.493	3.634	3.493	2.478	2.903
AdaBo-RW-U	2.321	3.776	3.366	2.493	3.045
AdaBo-RS-P	2.373	3.284	3.060	2.724	3.560
AdaBo-RS-U	2.351	3.201	3.067	2.866	3.515
MulBo-RW-P	2.201	3.813	3.612	2.478	2.896
MulBo-RW-U	2.276	3.836	3.485	2.448	2.955
MulBo-RS-P	2.254	3.537	3.075	2.746	3.388
MulBo-RS-U	2.299	3.485	2.903	2.776	3.537
LogBo-RW-P	1.649	1.530	2.821		
LogBo-RW-U	1.485	1.993	2.522		
LogBo-RS-P	2.037	1.522	2.440		
LogBo-RS-U	1.970	1.627	2.403		
RotFor-P	2.134	2.343		2.201	3.321
RotFor-U	2.313	2.037		2.455	3.194

The number of necessary rotation matrices, using ensembles of 100 classifiers, in OVO-E is 495000, while E-OVO only uses 100. Although these matrices are very sparse, they are too many for OVO-E with the current implementation. From the considered options, the worst one is E-OVO. The best option is not clear: according to the average performances (Tables 3 and 4) the best option is E but for RotFor-U the configuration OVA-E is better according to Tables 5, 6, 7 and 8.

4   Conclusion and Future Work

The performance of the combinations of binarization techniques and ensemble methods, using decision trees as base classifiers, over 67 data sets have been studied. As binarization techniques, the two most well-known were used: one vs. all and one vs. one. The used ensemble methods were Bagging, Random Subspaces, Random Forest, AdaBoost.M1, MultiBoost, LogitBoost and Rotation Forest.

One conclusion is that although binarization techniques can also improve the performance of ensemble methods, this improvement is not so clear as when

Table 8. Average ranks for kappa.

	E	OVA-E	OVO-E	E-OVA	E-OVO
Tree-P	1.761	2.507	1.731		
Tree-U	1.873	2.313	1.813		
Bagg-P	2.948	1.963	4.052	2.052	3.985
Bagg-U	2.993	1.851	4.127	2.060	3.970
RndSub-P	2.119	2.858	3.649	2.858	3.515
RndSub-U	2.313	2.679	3.754	2.604	3.649
RndFor	2.604	2.381	3.746	2.381	3.888
AdaBo-RW-P	2.425	3.560	3.537	2.552	2.925
AdaBo-RW-U	2.209	3.701	3.433	2.582	3.075
AdaBo-RS-P	2.291	3.313	3.007	2.843	3.545
AdaBo-RS-U	2.276	3.187	3.045	3.015	3.478
MulBo-RW-P	2.179	3.679	3.627	2.567	2.948
MulBo-RW-U	2.179	3.739	3.463	2.567	3.052
MulBo-RS-P	2.284	3.515	3.037	2.769	3.396
MulBo-RS-U	2.179	3.373	2.940	2.896	3.612
LogBo-RW-P	1.582	1.582	2.836		
LogBo-RW-U	1.500	1.948	2.552		
LogBo-RS-P	2.060	1.552	2.388		
LogBo-RS-U	2.015	1.575	2.410		
RotFor-P	2.082	2.381		2.239	3.299
RotFor-U	2.261	2.104		2.485	3.149

using classifiers that are not ensembles. This could be expected, is more difficult to improve better classifiers and ensembles are usually better than more simple classifiers. Besides, binarization techniques and ensembles can be considered similar, both combine several classifiers, and hence the improvements that can be obtained with one of them are already obtained with the other when they are combined.

Another conclusion is that although in previously reported comparisons usually OVO has better results than OVA, the situation is reversed with ensembles. This agrees with the idea that OVA is competitive when combining good classifiers. According to [23], OVA is as accurate as other binarization methods if the classifiers are "well-tuned". One way of improving classifiers is tuning its parameters, another is using ensembles.

When combining binarization techniques with ensembles, one question is the order of the techniques. In general, it is better to have ensembles of binarizers (OV?-E) than binarizers of ensembles (E-OV?), but for boosting variants the situation is the opposite.

This work has only considered the most basic binarization techniques. Other techniques could have better behaviour for ensembles. For instance, OVO can be used with ternary classifiers [1], OVO and OVA can be combined [19], the classifiers in OVO can be combined using a distance-based combination strategy [18], the classifiers in OVA can be combined with complementary classifiers [25]... Another family of binarization strategies than can be combined with ensemble methods are Error Correcting Output Codes (ECOC) [8,10,28].

An issue that has not been considered in this paper is the possible imbalance in the data sets. The methods considered in this paper could be evaluated on these data sets, using performance measures appropriate for this kind of data. The application of binarization techniques and other approaches for imbalanced problems have been studied in [11], these approaches could be also combined with ensemble methods. Moreover, there are ensemble methods for imbalanced data, some of them for multiclass [26]. These ensembles could be included in the study.

Acknowledgments. This work was partially supported by the project TIN2011-24046 of the Spanish Ministry of Economy and Competitiveness.

References

1. Angulo, C., Parra, X., Catala, A.: K-SVCR: a support vector machine for multiclass classification. Neurocomputing **55**(1), 57–77 (2003)
2. Bache, K., Lichman, M.: UCI machine learning repository (2013). http://archive.ics.uci.edu/ml
3. Bagheri, M.A., Gao, Q., Escalera, S.: A framework towards the unification of ensemble classification methods. In: 2013 12th International Conference on Machine Learning and Applications (ICMLA), vol. 2, pp. 351–355. IEEE, December 2013
4. Breiman, L.: Bagging predictors. Mach. Learn. **24**(2), 123–140 (1996)
5. Breiman, L.: Random forests. Mach. Learn. **45**(1), 5–32 (2001)
6. Demšar, J.: Statistical comparisons of classifiers over multiple data sets. J. Mach. Learn. Res. **7**, 1–30 (2006)
7. Dietterich, T.G.: Approximate statistical test for comparing supervised classification learning algorithms. Neural Comput. **10**(7), 1895–1923 (1998)
8. Dietterich, T.G., Bakiri, G.: Solving multiclass learning problems via error-correcting output codes. J. Artif. Intell. Res. **2**, 263–286 (1995)
9. Elomaa, T., Kääriäinen, M.: An analysis of reduced error pruning. J. Artif. Intell. Res. **15**, 163–187 (2001)
10. Escalera, S., Pujol, O., Radeva, P.: On the decoding process in ternary error-correcting output codes. IEEE Trans. Pattern Anal. Mach. Intell. **32**(1), 120–134 (2010)
11. Fernández, A., López, V., Galar, M., del Jesus, M.J., Herrera, F.: Analysing the classification of imbalanced data-sets with multiple classes: binarization techniques and ad-hoc approaches. Knowl.-Based Syst. **42**, 97–110 (2013)
12. Fernández-Delgado, M., Cernadas, E., Barro, S., Amorim, D.: Do we need hundreds of classifiers to solve real world classification problems? J. Mach. Learn. Res. **15**, 3133–3181 (2014)

13. Freund, Y., Schapire, R.E.: Experiments with a new boosting algorithm. In: 13th International Conference on Machine Learning, pp. 148–156. Morgan Kaufmann, San Francisco (1996)
14. Freund, Y., Schapire, R.E.: A decision-theoretic generalization of on-line learning and an application to boosting. J. Comput. Syst. Sci. **55**(1), 119–139 (1997)
15. Friedman, J., Hastie, T., Tibshirani, R.: Additive logistic regression: a statistical view of boosting. Ann. Stat. **95**(2), 337–407 (2000)
16. Fürnkranz, J.: Round robin ensembles. Intell. Data Anal. **7**(5), 385–403 (2003)
17. Galar, M., Fernández, A., Barrenechea, E., Bustince, H., Herrera, F.: An overview of ensemble methods for binary classifiers in multi-class problems: experimental study on one-vs-one and one-vs-all schemes. Pattern Recogn. **44**(8), 1761–1776 (2011)
18. Galar, M., Fernández, A., Barrenechea, E., Herrera, F.: DRCW-OVO: distance-based relative competence weighting combination for one-vs-one strategy in multi-class problems. Pattern Recogn. **48**, 28–42 (2014)
19. Garcia-Pedrajas, N., Ortiz-Boyer, D.: Improving multiclass pattern recognition by the combination of two strategies. IEEE Trans. Pattern Anal. Mach. Intell. **28**(6), 1001–1006 (2006)
20. Hall, M., Frank, E., Holmes, G., Pfahringer, B., Reutemann, P., Witten, I.H.: The WEKA data mining software: an update. SIGKDD Explor. **11**(1), 10–18 (2009)
21. Ho, T.K.: The random subspace method for constructing decision forests. IEEE Trans. Pattern Anal. Mach. Intell. **20**(8), 832–844 (1998)
22. Quinlan, J.R.: C4.5: Programs for Machine Learning. Morgan Kaufmann, San Mateo (1993). Machine Learning
23. Rifkin, R., Klautau, A.: In defense of one-vs-all classification. J. Mach. Learn. Res. **5**, 101–141 (2004)
24. Rodríguez, J.J., Kuncheva, L.I., Alonso, C.J.: Rotation forest: a new classifier ensemble method. IEEE Trans. Pattern Anal. Mach. Intell. **28**(10), 1619–1630 (2006)
25. Sesmero, M.P., Alonso-Weber, J.M., Gutierrez, G., Ledezma, A., Sanchis, A.: An ensemble approach of dual base learners for multi-class classification problems. Inf. Fusion **24**, 122–136 (2015)
26. Wang, S., Yao, X.: Multiclass imbalance problems: analysis and potential solutions. IEEE Trans. Syst. Man Cybern. Part B. Cybern. **42**(4), 1119–1130 (2012)
27. Webb, G.I.: Multiboosting: a technique for combining boosting and wagging. Mach. Learn. **40**(2), 159–196 (2000)
28. Windeatt, T., Ghaderi, R.: Coding and decoding strategies for multi-class learning problems. Inf. Fusion **4**(1), 11–21 (2003)
29. Zhou, Z.-H.: Ensemble Methods: Foundations and Algorithms. CRC Press, Boca Raton (2012)

Decision Tree-Based Multiple Classifier Systems: An FPGA Perspective

Mario Barbareschi, Salvatore Del Prete, Francesco Gargiulo,
Antonino Mazzeo, and Carlo Sansone$^{(\boxtimes)}$

Dipartimento di Ingegneria Elettrica e delle Tecnologie dell'Informazione (DIETI),
University of Naples Federico II, Via Claudio 21, 80125 Naples, Italy
{mario.barbareschi,salvatore.delprete,francesco.grg,
antonino.mazzeo,carlo.sansone}@unina.it

Abstract. Combining a hardware approach with a multiple classifier method can deeply improve system performance, since the multiple classifier system can successfully enhance the classification accuracy with respect to a single classifier, and a hardware implementation would lead to systems able to classify samples with high throughput and with a short latency. To the best of our knowledge, no paper in the literature takes into account the multiple classifier scheme as additional design parameter, mainly because of lack of efficient hardware combiner architecture.

In order to fill this gap, in this paper we will first propose a novel approach for an efficient hardware implementation of the majority voting combining rule. Then, we will illustrate a design methodology to suitably embed in a digital device a multiple classifier system having Decision Trees as base classifiers and a majority voting rule as combiner. Bagging, Boosting and Random Forests will be taken into account. We will prove the effectiveness of the proposed approach on two real case studies related to Big Data issues.

Keywords: Multiple classifier systems · Decision Tree · Bagging · Boosting · Random Forest · Field Programmable Gate Array

1 Introduction

Modern applications based on data analysis have been bringing new architectural design challenges. They define in the literature a new class of problems, addressed as *Big Data*, whose characteristics are grouped in the 5 'V's (Volume, Velocity, Variety, Veracity and Value), in order to indicate the inadequacy of the current computer technologies and design techniques when, at some point in time, these 'V's are increased to an unprecedented level. In particular, in the case of data classification, machine learning and pattern recognition algorithms have to deal with large data sets and with a very high number of samples per time unit (throughput) that have to be classified. Typical examples of problems with these characteristics are in the fields of intrusion detection [12], spam detection [10] and network traffic classification [4].

© Springer International Publishing Switzerland 2015
F. Schwenker et al. (Eds.): MCS 2015, LNCS 9132, pp. 194–205, 2015.
DOI: 10.1007/978-3-319-20248-8_17

The research community has made a big effort in devising not only new learning algorithms to enhance classification accuracy but also new design techniques, whose aim is to improve the classification speed, mainly exploiting hardware implementations. In the latter context, reconfigurable hardware technology, such as the Field Programmable Gate Array (FPGA), is able to realize high parallel, high speed and large digital designs, and, as it provides a programming flow that is somewhat similar to the software deploying, it allows easy and feasible hardware updates. These technological features are a promising solution for data classification tasks when a very high throughput value is required.

Among the multitude of different classification approaches proposed so far, Decision Trees (DTs) are one of the most suited for a hardware implementation, since they do not require arithmetic calculations, which are expensive to be realized, but only comparisons. The authors of [13] illustrated a high throughput DT classifier hardware accelerator design, mainly based on the pipeline technique, which reaches up to 114 times speed-up, compared with a software approach. With the aim of comparing power consumption of DT hardware and software approaches, authors of [7] introduced a methodology flow and, as result, they showed that the hardware version need only 0.03 % of the energy used by the software. In [1] a hardware accelerator for the DT detailed implementation is given, accomplished by exploiting a speculative approach on the node evaluation, reaching a very high throughput value, while in [2] same authors introduced a methodological flow to automatically obtain such hardware.

On the other hand, in many pattern recognition applications achieving an acceptable accuracy is conditioned by the large pattern variability, whose distribution cannot be simply modelled. This affects the results of the classification system so that, once this has been designed, its performance cannot be improved beyond a certain bound, despite efforts at refining either the classification or the description method [9]. A possible solution is the use of a multiple classifier system: the consensus of a set of classifiers may compensate for the weakness of a single classifier.

Combining hardware accelerators with multiple classifier techniques can dramatically improve the system performance, as the multiple classifier systems are able to successfully enhance the classification accuracy and designs realized in hardware perform classification of samples with really high throughput and short latency. To the best of our knowledge, no paper in the literature takes into account the multiple classifier scheme as an additional design parameter, mainly because of lack of efficient hardware combiner architectures.

For those reasons and starting from the previous considerations, in this paper we try to fill the gap by presenting an efficient hardware implementation of the majority voting rule. Moreover, we illustrate a design methodology to suitably embed in a digital device a multiple classifier system, having a DT as base classifier and a majority voting rule as combiner. In particular, Bagging, Boosting and Random Forests [11] have been considered as multiple classifier systems. By taking into account the constraints given by the specific problem, the proposed methodology is able to select the best possible hardware multiple classifier system by considering classification accuracy, throughput and hardware latency. In order to avoid the generation of unfeasible ensembles (i.e., they cannot be synthesized in hardware), we also present an early prediction approach to do a preliminary estimate of the required hardware resources, empirically exploiting measurements and adopting as base classifier the hardware version of the DT introduced in [1] and as hardware a Xilinx Virtex 5 FPGA device.

The paper is structured as follows: in Sect. 2 we briefly introduce the hardware DT classifier and the combiner that we exploit to implement a hardware multiple classifier system. Section 3 contains a detailed description of the proposed design methodology. Thus, in Sect. 4, we demonstrate the effectiveness of the proposed approach through two real case studies, namely spam detection and IP traffic classification, discussing the main performance aspects. At the end, the Sect. 5 concludes the paper.

2   From Classification Model to Hardware Accelerator

In order to gather useful data about the performance of hardware classifiers and analyse them by varying the classification parameters, it is mandatory to have what is closest to a physical realization of the hardware components under test, such as the description at the *Place and Route* (PAR) level. Essentially, a PAR description has a fine grain level of physical details as it describes a digital circuit in terms of what will be realized on the technological target. For instance, a PAR description for a Xilinx FPGA contains configuration and allocation of the Look-up Tables (LUTs), Registers, Slices and routing resources involved into the design.

Towards this aim, experiments need for designs described as HDL projects, which implement the hardware accelerator for a specific classification module.

2.1   Decision Tree Implemented on FPGA

Basically, the hardware implementation for a multiple classifier scheme is directly inherited from the model structure. As the multiple classifier model combines classification techniques, which are not dependent one another during the evaluation phase, and makes their predictions with a combining rule, the hardware structure is designed with parallel classification entities that execute high speed classification in parallel and with a hardware combiner which quickly organizes all the classifiers' outputs. In this paper we consider the DT as base classifier model, since it can be successfully implemented in hardware and it is suitable in a wide range of applications and domains. In particular, for the proposed multiple classifier architecture, we exploit the hardware accelerator presented in [1] because it is specifically designed for a FPGA technology, as it deeply exploits the FPGAs' parallel structure. Moreover, the authors have deployed a tool whereby hardware models can be automatically generated from formal models described in Predictor Mark-up Model Language (PMML), which represents a standard exploitable as output artefact by a wide range of analytic frameworks (e.g. KNIME[1] or WEKA[2]) [2].

The DT model mapping is accomplished by implementing each tree node as a binary comparator, which, once received the feature value, returns a boolean value. In order to exploit intrinsic hardware parallelism, all the tree nodes work in parallel and their decision values feed the boolean network that computes which tree leaf has been reached. In particular, the boolean network needs as much input as the DT nodes and gives outputs equal to the number of classes, such that only one output is high each time. This approach turns out to be speculative, since the DT model does not consider all the decisions at the same time, but only the ones that belong to a single path, since the visiting algorithm traverses the decision nodes one by one accordance with the comparison results. The speculation implies very fast computation since there are no dependencies between the decisions that can be simultaneously evaluated.

In order to support a multiple DT hardware accelerator, in the next subsection we present the majority voting rule as a hardware combiner. Furthermore, for extending the automatic hardware description generation illustrated in [2], we have successfully integrated the automatic generation of such a combiner in the previously developed tool *PMML2VHDL*.

2.2   Hardware Combiner

In hardware, the combiner of a multiple classifier system is one of the most influential elements both for latency and area occupation.

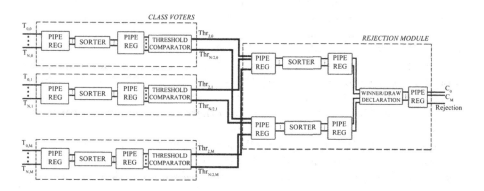

Fig. 1. Majority voter implemented as a pipelined odd-even sorter, which also embeds a rejection module.

According to its implementation, the combiner could be a bottleneck, so it is necessary to find a balanced design which can be a good trade-off between resource occupation and maximum throughput. Among possible design solutions, in this paper we adopt the one based on the *winning threshold* and on the *sorting network*. First of all, the idea is to exploit a combiner that does not care about which class is voted by which tree, but only how many votes a class gets. Each DT expresses its own decision as a decoded output, hence only one bit is high, and each class gets a certain amount of votes, which corresponds to the number of high bits received by each DT. Rather than use a binary adder to count high bits, it is simpler to collect the votes in a vector by shifting all the high bits at the beginning of it and verifying if they are enough to declare someone as the winner.

As depicted in Fig. 1, we design the combiner as an odd-even sorter for each class, which is a component that implements a simple sorting algorithm closely related to bubble-sort, but in a parallel version. Indeed, the architecture compares all the couples of adjacent elements, whose first member occupies an odd position in the vector, swapping them if they are in the wrong order. Then, it repeats the operation for even-indexed couples. The whole process is iterated until the vector is totally ordered. In hardware, the algorithm is implemented as a pipelined sorting network, such that the groups of 2-selectors work in parallel reaching very high speed. The ordered vector is the input of another component, which verifies whether a threshold of votes is reached according to the product of the first X values in the vector, where X is the threshold value. The outputs of all these components are compared in the rejection module, i.e. another sorting network, in order to declare a winner or a draw situation.

2.3   Automatic Generator of Classification Models

Since our goal is to control the classification models' characteristics, such as number of nodes, maximum tree depth, number of classes, number of trees and so on, in order to evaluate their influence on hardware classifier performance, we develop *PMMLGen*, namely a Java tool which builds PMML models with desired parameters. This application allows us to avoid learning for actual training sets to obtain classification models (i.e., DTs) with the desired characteristics, as it can automatically build working PMML models being able to control some tree-based classifiers parameters, while the others are randomly and coherently picked.

Exploiting *PMMLGen*, we have collected some models whose characteristics are needed for defining trends of the hardware-related parameters according to their features, and hence for estimate an *early prediction* function, as detailed in the next Section. This is useful to reduce test cases, since it avoids generation of unfeasible ensembles, i.e. models that cannot be synthesized in a specific targeted hardware.

3   Methodology for Performance Evaluation: Early Prediction Function

As the space of possible classification system solutions is very large, in this Section we introduce an area-occupation Early Prediction (EP) function that is useful to early discard models which likely lead to designs that are not feasible to be implemented in the target device, which in this paper is an FPGA. To this aim, we have obtained such a function through a step-wise regression applied on the results of a test-suite made up of artificial ensembles generated by *PMMLGen*. We have focused on the effect of each parameter, varying them one per time and keeping the others fixed at a specific value. In particular, we have considered: number of nodes for each tree and, consequently, the overall number of nodes in the ensemble (*#Nodes*); number of trees (*#Trees*); number of classes (*#Classes*) and number of features (*#Features*). Therefore, we can give a general form for the *EP*:

$$EP = f(\#Classes, \#Features, \#Trees, \#Nodes). \tag{1}$$

The actual expression of *EP* strictly depends on the device considered, since it refers to its technological characteristics.

Even if the *EP* should be used to predict, from the parameters of the obtained model, whether the multiple classifier system requires an amount of resources suitable for the targeted device, it might be useful even without having trained models. The first three parameters are clearly defined by the problem at hand, but the last one is tightly coupled with the dataset and with the learning algorithm. To preliminarily have a suitable estimation about the number of nodes which will likely characterize the trained models, it could be enough to get only one complete training for a significant ensemble model case, e.g. on a small ensemble with 5 trees. In this way, the *#Nodes* parameter of the *EP* function can be estimated as the averaged number of trees nodes, since the relation between the number of nodes and number of trees is quite linear in most cases. Therefore we claim that this preliminary estimation of the nodes is an overestimation since, with the growing of the involved trees, the number of nodes might linearly grow or be constant.

Performance Function. Once the number of possible design solutions have been reduced, selected classifier systems have to be implemented in order to get information about their accuracy, latency and throughput values. As stated before, this step involves hardware synthesis tools which translate designs from a PAR description.

As for the accuracy, we adopt a *k*-fold cross validation technique for each classification model under test. Having the performance values for each feasible design, it is possible to assign a *performance value* to each involved model and, in the end, select the best one according to the requirements of the application. Towards this aim, we define a suitable performance function (*P*) assuming a linear dependence on *accuracy*, *latency* and *throughput*, hence its expression is given by:

$$P = \alpha \cdot Accuracy^{Norm} + \beta \cdot \frac{1}{Latency}^{Norm} + \gamma \cdot Throughput^{Norm}. \tag{2}$$

As one can notice, since *accuracy*, *latency* and *throughput* have different ranges and measurement units, there is the need for the value normalization.

An effective solution could be the evaluation of the performance improvement when a multiple classifier system is used, with respect to the use of its base classifier, i.e. the single DT. Hence, in Eq. 2, the notation $<v>^{Norm}$ stands for adopting the following normalization rule:

$$<v>^{Norm} \Rightarrow \frac{<v> - <v>_{DT}}{<v>_{DT}} \tag{3}$$

where $<v>_{DT}$ represents the value from a single DT built on all the data.

In conclusion, we can also observe that, in most cases, *latency* and *throughput* are not both critical, so that one of the weights of Eq. 2 can be considered null and the remaining two can be fixed to α and $(1-\alpha)$, respectively. In this specific case, given α, the design which maximizes P can be implemented in hardware.

4   Experimental Results

The goal of this section is twofold: first of all we test the proposed methodology, showing how the *early prediction* function can estimate the maximum number of trees according to the chosen ensemble strategy; then, exploiting the *performance function P*, we show the best classification system to be implemented in hardware.

For the latter aim, we prove the effectiveness of the previously introduced methodology with two case studies: the first one refers to spam detection, characterized by a huge amount of data that must be typically processed in a fixed time. The second one is related to the classification of IP traffic traces: in this case the main constraint to be satisfied is real-time classification [5,6]. Moreover, while in the former case we have a binary problem with dozens of features, in the second we consider a multi-class problem with few features.

For all the tests, we considered as target reference the hardware platform Xilinx Virtex 5 XC5VLX110T, whose characteristics are illustrated in Table 1.

Table 1. Xilinx Virtex 5 XC5VLX110T characteristics

Array (Row x Col)	Slices	Slice registers	Slice LUTs	Total I/O banks	Max user I/OBs
160×54	17280	69120	69120	20	640

4.1   Early Prediction

For defining the EP function for the hardware device we are considering, we generated several artificial ensembles by using *PMMLGen* with model parameters whose values were in the ranges detailed in Table 2.

We trained a single DT classifier, as well as the Bagging, AdaBoost and Random Forest multiple classifier systems, by using the KNIME Analytic framework. Once obtained the artificially generated ensembles, were saved as PMML files

Table 2. Value ranges of the parameters.

Parameter	Lower Bound	Upper Bound
#$Trees$	5	50
#$Classes$	4	32
#$Features$	4	32
#$Nodes$	55	3375

and later translated in VHDL by using the *PMML2VHDL* framework in order to synthesize them in hardware and retrieve their performance characteristics. For the Xilinx Virtex-5 we found the following expression of the *EP* function:

$$EP = 13 * \#Classes + 33 * \#Features - 74 * \#Trees + 9 * \#Nodes^*.$$

where #$Nodes^*$ is an estimation of the actual number of nodes in the ensemble, made as described in the previous Section.

In Fig. 2 we report the obtained *EP* function. The estimation quality is globally good as the real required hardware resources are really close to the predicted ones; hence, *EP* can be used to evaluate the maximum feasible number of trees for each ensemble strategy.

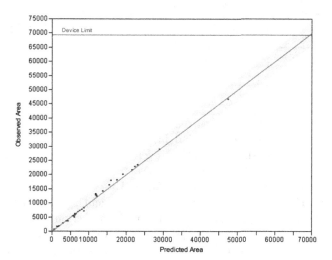

Fig. 2. Predicted vs Observed plot of Early Prediction function; the red line represents the hardware limit for the Xilinx Virtex 5 XC5VLX110T board.

In the following subsections we use the early prediction function and, for each classification system, we consider the accuracy evaluated by means of a 10-fold cross validation strategy.

Table 3. SPAM e-mail detection - VHDL Classifiers. Best solutions in terms of P are reported in bold.

(a) Decision Tree

#VHDL Nodes	Max Depth	Throughput [MS/s]	Accuracy [%]	P $\alpha = 0.25$	P $\alpha = 0.50$	P $\alpha = 0.75$	P $\alpha = 1.00$
206	41	112.8796	91.806	0.000	0.000	0.000	0.000

(b) Bagging

#Trees	#VHDL Nodes	Max Depth	Throughput [MS/s]	Accuracy [%]	P $\alpha = 0.25$	P $\alpha = 0.50$	P $\alpha = 0.75$	P $\alpha = 1.00$
5	276	28	123.09	92.13	0.069	0.047	0.025	0.004
10	296	21	129.43	91.85	0.110	0.074	0.037	0.001
15	281	18	130.40	91.26	0.115	0.075	0.034	-0.006
20	282	19	126.29	91.66	0.089	0.059	0.029	-0.002
25	266	14	133.56	90.85	**0.135**	**0.086**	0.038	-0.010
50	312	11	125.75	91.38	0.084	0.055	0.025	-0.005
100	348	9	127.15	90.89	0.092	0.058	0.024	-0.010
125	364	7	131.13	90.29	0.117	0.073	0.028	-0.017

(c) Random Forest

#Trees	#VHDL Nodes	Max Depth	Throughput [MS/s]	Accuracy [%]	P $\alpha = 0.25$	P $\alpha = 0.50$	P $\alpha = 0.75$	P $\alpha = 1.00$
5	1536	45	125.33	94.00	0.089	0.067	0.046	0.024
10	3027	45	124.27	94.88	0.084	0.067	0.050	0.033
15	4496	45	113.86	95.11	0.015	0.022	0.029	0.036
20	5939	45	120.85	95.33	0.063	0.054	0.046	0.038
24	6956	43	126.28	95.81	0.100	0.081	**0.062**	**0.044**

(d) Boosting

#Trees	#VHDL Nodes	Max Depth	Throughput [MS/s]	Accuracy [%]	P $\alpha = 0.25$	P $\alpha = 0.50$	P $\alpha = 0.75$	P $\alpha = 1.00$
5	1210	59	127.83	94.02	0.105	0.078	0.051	0.024
10	2657	60	100.27	95.30	-0.074	-0.037	0.001	0.038
15	3813	60	107.82	92.96	-0.031	-0.016	-0.002	0.013
20	5109	61	119.59	94.48	0.052	0.044	0.037	0.029
30	6764	62	112.89	94.05	0.006	0.012	0.018	0.024

4.2 Spam Detection

For this case study we used the *Spambase* dataset, publicly available on the UCI repository[3]. This dataset contains 4601 instances (1813 Spam cases) characterized by 57 continuous features. Note that, even if the number of training samples is not so significant, it is likely that a spam detection system should process a huge amount of data when operating in the field.

In order to select the best solution for this particular problem, we used the *performance function* P introduced in the previous Section. Since, in general, spam detection does not need to be performed in real-time, we are not particularly interested in minimizing *latency*, while we need to maximize *throughput*, since it is important to classify as many email as possible in a given time unit. So we can set $\beta = 0$ and $\gamma = 1 - \alpha$, thus re-writing the *performance function* as:

$$P = \alpha \cdot Accuracy^{Norm} + (1 - \alpha) \cdot Throughput^{Norm}.$$

We considered four possible values for α, i.e., $\alpha = \{0.25; 0.50; 0.75; 1.00\}$ in order to differently weight *accuracy* and *throughput*. Note that the last value

---

[3] https://archive.ics.uci.edu/ml/datasets/Spambase.

Table 4. Internet Traffic Classification - VHDL Classifiers. Best solutions in terms of P are reported in bold.

(a) Decision Tree

#VHDL Nodes	Max Depth	Latency [ns]	Accuracy [%]	P $\alpha = 0.25$	P $\alpha = 0.50$	P $\alpha = 0.75$	P $\alpha = 1.00$
111	11	49.98	91.78	0.000	0.000	0.000	0.000

(b) Bagging

#Trees	Max Depth	Latency [ns]	Accuracy [%]	P $\alpha = 0.25$	P $\alpha = 0.50$	P $\alpha = 0.75$	P $\alpha = 1.00$
5	11	91.66	92.96	-0.338	-0.221	-0.104	**0.013**
10	10	79.31	92.54	-0.275	-0.181	-0.086	0.008
30	10	84.77	91.79	-0.308	-0.205	-0.103	0.00
60	10	81.98	91.15	-0.294	-0.231	-0.119	-0.007
90	10	86.43	90.88	-0.319	-0.216	-0.113	-0.010

(c) Random Forest

#Trees	Max Depth	Latency [ns]	Accuracy [%]	P $\alpha = 0.25$	P $\alpha = 0.50$	P $\alpha = 0.75$	P $\alpha = 1.00$
5	13	85.00	91.85	-0.309	-0.206	-0.102	0.001
10	13	95.76	92.05	-0.358	-0.238	-0.117	0.003
20	13	96.22	92.19	-0.359	-0.238	-0.117	0.004
30	13	92.32	92.20	-0.343	-0.227	-0.111	0.005
35	14	96.99	92.21	-0.362	-0.240	-0.118	0.005

(d) Boosting

#Trees	Max Depth	Latency [ns]	Accuracy [%]	P $\alpha = 0.25$	P $\alpha = 0.50$	P $\alpha = 0.75$	P $\alpha = 1.00$
5	14	87.92	92.46	-0.322	-0.212	-0.102	0.007
10	13	87.20	92.33	-0.319	-0.210	-0.102	0.006
25	15	96.04	92.31	-0.358	-0.237	-0.116	0.006
50	14	98.15	92.18	-0.367	-0.243	-0.119	0.004
75	15	94.53	92.22	-0.325	-0.233	-0.114	0.005

corresponds to the case in which only accuracy is considered, i.e. we are searching for the system with the best accuracy that can be implemented in hardware.

Using the *EP* function to estimate the required resources in terms of area on the FPGA board, we obtained the maximum feasible number of trees for each ensemble strategy, i.e. Bagging (158), Random Forest (24) and Boosting (30).

The results reported in Table 3 show that if we want to take care of the *throughput* ($\alpha = 0.25$) or prefer to equally weight *accuracy* and *throughput* ($\alpha = 0.5$), we have to choose the Bagging algorithm with 25 trees. On the other hand, if we believe that accuracy is more important ($\alpha \geq 0.75$), Random Forests with 24 base classifiers should be chosen. In all cases, we have several multiple classifier systems whose overall performance (according to our definition) is better than the one obtained by the single DT.

4.3   Traffic Classification

In this case we used a dataset provided by the Lawrence Berkeley National Laboratory (LBNL)[4], already used in [3,8]. The dataset is composed by 134246 samples, characterized by 7 continuous features and 6 different classes: *POP3, FTP, SMTP, HTTP, BIT-TORRENT, MSN*. Again, we followed the above presented methodology in order to select the best possible hardware solution for this

---

[4] http://ee.lbl.gov/anonymized-traces.html.

problem. Since, unlike spam detection, traffic classification would need to be performed in real-time, we surely need to minimize *latency*, while we would not be particularly interested in maximizing *throughput*. So we can simplify the expression of the *performance function*, by setting $\gamma = 0$ and $\beta = 1 - \alpha$ as in the following: $P = \alpha \cdot Accuracy^{Norm} + (1 - \alpha) \cdot \frac{1}{Latency}^{Norm}$.

By using the EP function we obtained the maximum feasible number of trees for each ensemble algorithm, i.e. Bagging (95), Random Forest (37) and Boosting (77).

Then, according to these results, we are able to select the best solutions for the problem at hand. As in the previous case, we considered four possible values for α, i.e., $\alpha = \{0.25; 0.50; 0.75; 1.00\}$. In this case it is interesting to note that, since the DT has a very small *latency* time, the single classifier solution should be chosen according to our *performance function*, when we want to use the *latency* parameter. On the other hand, if we are only interested in maximizing *accuracy*, a bagging ensemble with 5 trees should be implemented in hardware (Table 4).

5   Conclusion

In this paper we presented a novel approach for an efficient hardware implementation of the majority voting combining rule and illustrated a design methodology to suitably embed in a digital device a multiple classifier system, having a DT as base classifier and a majority voting rule as combiner. Bagging, boosting and random forests have been considered as multiple classifier systems. Taking into account the constraints given by the specific problem, in the proposed methodology we introduced a performance function P that was able to select the best possible hardware classifier system by considering classification accuracy, throughput and hardware latency. We also presented an early prediction EP function to preliminary estimate the number of trees usable within the ensembles approaches according to the hardware constraints.

We presented the results of the proposed approach by considering two different problems: spam detection, a binary classification problem where both throughput and accuracy need to be maximized and the classification of IP traffic traces, a multi-class problem where latency need to be minimized, while preserving an high accuracy. In both cases, we considered different scenarios, by weighting in different ways the contribution of throughput (or latency) and accuracy to the overall system performance, by varying a suitably defined parameter.

In case of spam detection, we found that there was always a multiple classifier system that outperforms the single DT classifier, as could be expected. On the other hand, when the traffic classification problem was addressed, it happened that a single classifier solution is the most suitable one for several scenarios. The multiple classifier approach in this case should be preferred for an hardware implementation if we are only interested in maximizing accuracy.

As future work we are planning to investigate the possibility of extending our methodology to other multiple classifier systems.

Acknowledgments. The research leading to these results has been partially supported by the RoDyMan project, which has received funding from the European Research Council (FP7 IDEAS) under Advanced Grant agreement number 320992. The authors are solely responsible for its content. It does not represent the opinion of the European Community and the Community is not responsible for any use that might be made of the information contained therein.

References

1. Amato, F., Barbareschi, M., Casola, V., Mazzeo, A.: An fpga-based smart classifier for decision support systems. In: Zavoral, F., Jung, J.J., Badica, C. (eds.) Intelligent Distributed Computing VII, pp. 289–299. Springer, Switzerland (2014)
2. Amato, F., Barbareschi, M., Casola, V., Mazzeo, A., Romano, S.: Towards automatic generation of hardware classifiers. In: Aversa, R., Kołodziej, J., Zhang, J., Amato, F., Fortino, G. (eds.) ICA3PP 2013, Part II. LNCS, vol. 8286, pp. 125–132. Springer, Heidelberg (2013)
3. Dainotti, A., Gargiulo, F., Kuncheva, L.I., Pescapè, A., Sansone, C.: Identification of traffic flows hiding behind TCP port 80. In: ICC, pp. 1–6. IEEE (2010)
4. Dainotti, A., Pescapè, A., Claffy, K.C.: Issues and future directions in traffic classification. IEEE Netw. **26**(1), 35–40 (2012)
5. Dainotti, A., Pescapé, A., Sansone, C.: Early classification of network traffic through multi-classification. In: Domingo-Pascual, J., Shavitt, Y., Uhlig, S. (eds.) TMA 2011. LNCS, vol. 6613, pp. 122–135. Springer, Heidelberg (2011)
6. Dainotti, A., Pescapé, A., Sansone, C., Quintavalle, A.: Using a behaviour knowledge space approach for detecting unknown IP traffic flows. In: Sansone, C., Kittler, J., Roli, F. (eds.) MCS 2011. LNCS, vol. 6713, pp. 360–369. Springer, Heidelberg (2011)
7. Franca, A.L.P.d., Jasinski, R.P., Pedroni, V.A., Santin, A.O.: Moving network protection from software to hardware: an energy efficiency analysis. In: 2014 IEEE Computer Society Annual Symposium on VLSI, pp. 456–461. IEEE, Tampa (2014)
8. Gargiulo, F., Kuncheva, L.I., Sansone, C.: Network protocol verification by a classifier selection ensemble. In: Benediktsson, J.A., Kittler, J., Roli, F. (eds.) MCS 2009. LNCS, vol. 5519, pp. 314–323. Springer, Heidelberg (2009)
9. Gargiulo, F., Mazzariello, C., Sansone, C.: Multiple classifier systems: theory, applications and tools. In: Bianchini, M., Maggini, M., Jain, L.C. (eds.) Handbook on Neural Information Processing, pp. 335–378. Springer, Heidelberg (2013)
10. Guzella, T.S., Caminhas, W.M.: A review of machine learning approaches to spam filtering. Expert Syst. Appl. **36**(7), 10206–10222 (2009)
11. Kuncheva, L.: Combining Pattern Classifiers: Methods and Algorithms, 2nd edn. Wiley-Interscience, Hoboken (2014)
12. Mitchell, R., Chen, I.R.: A survey of intrusion detection techniques for cyber-physical systems. ACM Comput. Surv. **46**(4), 55:1–55:29 (2014)
13. Saqib, F., Dutta, A., Plusquellic, J., Ortiz, P., Pattichis, M.: Pipelined decision tree classification accelerator implementation in fpga (dt-caif). IEEE Trans. Comput. **64**(1), 280–285 (2015)

An Empirical Investigation on the Use of Diversity for Creation of Classifier Ensembles

Muhammad A.O. Ahmed, Luca Didaci, Giorgio Fumera$^{(\boxtimes)}$, and Fabio Roli

Department of Electrical and Electronic Engineering, University of Cagliari,
Piazza d'Armi, 09123 Cagliari, Italy
{muhammad.ahmed,didaci,fumera,roli}@diee.unica.it
http://pralab.diee.unica.it

Abstract. We address one of the main open issues about the use of diversity in multiple classifier systems: the effectiveness of the explicit use of diversity measures for creation of classifier ensembles. So far, diversity measures have been mostly used for ensemble pruning, namely, for selecting a subset of classifiers out of an original, larger ensemble. Here we focus on pruning techniques based on forward/backward selection, since they allow a direct comparison with the simple estimation of accuracy of classifier ensemble. We empirically carry out this comparison for several diversity measures and benchmark data sets, using bagging as the ensemble construction technique, and majority voting as the fusion rule. Our results provide further and more direct evidence to previous observations against the effectiveness of the use of diversity measures for ensemble pruning, but also show that, combined with ensemble accuracy estimated on a validation set, diversity can have a regularization effect when the validation set size is small.

Keywords: Diversity · Ensemble pruning · Forward/backward selection · Ensemble construction

1   Introduction

After about twenty years of active research in the classifier ensemble field, understanding the notion of diversity remains one of the main open problems [11,25]. On the one hand, there is a general agreement on the qualitative definition of diversity and on its role, e.g.: "it is desired that the individual learners should be *accurate and diverse*" [25]; "Common sense suggests that the classifiers in the ensemble should be as accurate as possible and should not make coincident errors" [11] (Chap. 8). On the other hand, measuring diversity and explicitly using it for ensemble construction exhibits several open issues.

A number of **diversity measures** have been proposed over the years [9,11,25]. Most measures have been derived intuitively, as attempts to formally characterize the pattern of individual classifiers' errors (e.g., the Double-Fault and Disagreement measures [11]). In particular, it has been clearly pointed out that diversity measures alone can not be monotonically related to ensemble accuracy, since the

© Springer International Publishing Switzerland 2015
F. Schwenker et al. (Eds.): MCS 2015, LNCS 9132, pp. 206–219, 2015.
DOI: 10.1007/978-3-319-20248-8_18

latter depends instead on a trade-off between diversity and individual classifiers' performance [11,19]; quoting from [11] (Chap. 8), looking for a diversity measure strongly related to ensemble performance runs the risk of "replacing a simple calculation of the ensemble error by a clumsy proxy which we call diversity." A few other measures have been inspired by *exact* error decompositions derived in the regression field, despite the lack of a direct analogy with regression problems was pointed out in [2]: the Kohavi-Wolpert Variance [9] (and our attempt in [6]) was inspired by the bias-variance-covariance error decomposition [21], and the measure derived in [3] (which we extended in [6]) by the ambiguity decomposition [8]. The rationale of such measures is to look for exact, additive decompositions of the ensemble error into terms accounting for individual classifiers' performance, and terms hopefully interpretable as diversity; the results of [3] provided useful insights, leading to the concept of "good" and "bad" diversity. Several authors also analyzed, empirically or analytically, the connection between ensemble performance on one side, and the pattern of individual classifiers' performance and existing diversity measures on the other side (e.g., [10,19]). Such a relationship turned out to be far from clear-cut, and no "right" diversity measure has emerged so far.

Almost all the existing methods that **explicitly use diversity for ensemble construction** follow the overproduce and choose approach (except for [24], where a diversity measure is used in an ensemble *learning* algorithm). It consists of first generating a large ensemble (e.g., using bagging) and then selecting the most accurate subset of classifiers (usually with a predefined size). This is known as ensemble *pruning, selection* or *thinning*. Since this problem has exponential complexity in the size of the original ensemble, several heuristics have been proposed. In this context, diversity measures have been used in the objective function of pruning methods, to look for a trade-off between individual classifiers' performance and diversity. The effectiveness of such an approach has however been questioned by several authors, based also on empirical evidences [11,19] (Chap. 8.3). In particular, its actual advantage over directly evaluating ensemble performance (estimated, e.g., from validation data) is not clear yet. On the other hand, it is well known that popular and effective ensemble construction techniques like bagging and boosting do not use any explicit diversity measure.

In [6] we discussed the above issues, focusing on the derivation of exact decompositions of the ensemble error, and outlined several research directions. One of them, which we start addressing in this work, consists of comparing the effectiveness of explicitly using diversity measures in ensemble pruning, with the simple estimation of ensemble performance. Although many pruning methods have been proposed so far, the above comparison has been carried out by only a few authors, and with a limited scope. In this work we focus on pruning methods based on forward/backward selection (FS/BS) algorithms, which are the easiest ones on which such a comparison can be made, and carry out an empirical investigation on 23 benchmark data sets, using the popular bagging as the ensemble construction technique, and majority voting as the fusion rule. We evaluate ten well known diversity measures analyzed in [9], and five measures specifically defined for ensemble pruning. We also evaluate the effect of the validation set size on ensemble pruning effectiveness.

Algorithm 1. Forward Selection algorithm for ensemble pruning

---

Input: an ensemble E of N classifiers; a desired ensemble size $L < N$; a validation set V; an objective function m (to be computed on V)

Output: a subset of L classifiers from E

$C \leftarrow$ the most accurate individual classifier from E

$S \leftarrow \{C\}$

for $k = 2, \ldots, L$ **do**

   $C^* \leftarrow \arg\max_{C \in E \setminus S} m(S \bigcup \{C\})$

   $S \leftarrow S \bigcup \{C^*\}$

end for

return S

---

2   Ensemble Pruning with Forward/Backward Selection

Ensemble pruning methods can be categorized as follows [20]:

- **Ranking-based:** individual classifiers are first ranked according to some criterion, and then the top-L are selected to form the final ensemble.
- **Clustering-based:** individual classifiers are first clustered based on the similarity of their predictions; each cluster is then pruned to remove redundant classifiers, and the remaining ones in each cluster are finally combined.
- **Optimization-based** methods search for a subset of the original ensemble that optimizes a given objective function, which can include a diversity measure. To avoid exhaustive search, three main heuristic search strategies have been proposed: hill climbing (often implemented as FS or BS), genetic algorithms, and semi-definite programming.

We focus on optimization-based methods in which FS/BS is used, since they allow a direct comparison between the simple estimation of ensemble accuracy and objective functions involving diversity. Several pruning methods based on FS/BS, together with specific objective functions, have been proposed so far, including [1,4,13–17]. Given an initial ensemble E of size N, FS constructs a pruned ensemble S of size $L < N$ by starting from the best individual classifier from E, and iteratively adding a classifier to S by maximizing a given objective function (see Algorithm 1).[1] The BS algorithm works similarly, iteratively removing from E one classifier at a time. More refined versions of FS/BS have also been proposed, which include a back-fitting step [13].

Three kinds of objective functions have been proposed so far:

- The ensemble performance, [13] (reduce-error pruning technique), [4,12].
- Diversity measures alone, disregarding the performance of individual classifiers and of the ensemble, [13] (Kullback-Leibler Divergence pruning), [17] and [1] (kappa-thinning).

---

[1] If no predefined size is given, FS stops when all the classifiers from E have been added, and returns the best ensemble among the N ones obtained at every iteration.

– Measures specifically defined for ensemble pruning. They combine into a single scalar the individual classifiers' performance and the *complementarity* between their errors [14–16] and [1] (AID thinning and Concurrency thinning). We will refer to them as *pruning measures*.

Among the existing pruning measures, we focus on the following ones. Let $(\mathbf{x}, y)$ denote a sample with its class label, V the validation set, E and S the original and the current pruned ensemble, C^* the candidate classifier to be added to (or removed from) S, and $S(\mathbf{x})$ the label assigned to $\mathbf{x}$ by S.

– A measure aimed at minimizing the number of coincident errors between ensemble members, when majority voting is used, to be used in the FS algorithm [16] (Sect. 5.2). It selects the classifier C^* that correctly labels the highest number of validation samples, among the ones misclassified by the majority of classifiers in the current ensemble S:

$$C^* = \arg\min_{C \in E \setminus S} \sum_{(\mathbf{x},y) \in V} \quad \begin{aligned} & I\left[C(\mathbf{x}) \neq y \wedge S(\mathbf{x}) \neq y\right] \\ & - I\left[C(\mathbf{x}) = y \wedge S(\mathbf{x}) \neq y\right], \end{aligned} \tag{1}$$

where $I[A] = 1$ if $A = $True, and $I[A] = 0$ otherwise.
– Two measures proposed in [14] to be used in the FS algorithm, with the majority voting rule: Complementariness (the sum of validation samples which are wrongly classified by the current ensemble, but not by the candidate classifier) and Margin Distance. The former is a variant of Eq. (1). They are respectively defined as:

$$C^* = \arg\max_{C \in E \setminus S} \sum_{(\mathbf{x},y) \in V} I\left[C(\mathbf{x}) = y \wedge S(\mathbf{x}) \neq y\right], \tag{2}$$

$$C^* = \arg\min_{C \in E \setminus S} \left\| \mathbf{o} - \frac{1}{|E|}\left(\mathbf{c}_C + \sum_{C' \in S} \mathbf{c}_{C'}\right) \right\|_2^2, \tag{3}$$

where $\mathbf{c}_{C'}$ is a $|V|$-dimensional vector whose i-th element is defined as:

$$2I[C'(\mathbf{x}_i) = y_i] - 1 \in \{-1, +1\},$$

and $\mathbf{o}$ is defined as a constant vector whose components are all identical to some value p, with $0 < p < 1$.
– A measure proposed in the context of the Concurrency thinning technique in [1], based on BS. It chooses the classifier to be removed from S with the aim of penalizing the agreement on correctly classified samples (this is a variant of Eq. (1) as well):

$$C^* = \arg\min_{C \in S} \sum_{(\mathbf{x},y) \in V} \quad \begin{aligned} & I\left[C(\mathbf{x}) = y \wedge S(\mathbf{x}) = y\right] \\ & + 2I\left[C(\mathbf{x}) = y \wedge S(\mathbf{x}) \neq y\right] \\ & - 2I\left[C(\mathbf{x}) \neq y \wedge S(\mathbf{x}) \neq y\right]. \end{aligned} \tag{4}$$

– The Uncertainty Weighted Accuracy (UWA), to be used in the FS algorithm; it was proposed in [15] as a variant of the Concurrency measure of Eq. (4):

$$
\begin{aligned}
C^* = \arg\max_{C \in E \setminus S} \sum_{(\mathbf{x},y) \in V} \quad & NF(\mathbf{x}) \times I\left[C(\mathbf{x}) = y \wedge S(\mathbf{x}) = y\right] \\
& + NT(\mathbf{x}) \times I\left[C(\mathbf{x}) = y \wedge S(\mathbf{x}) \neq y\right] \\
& - NF(\mathbf{x}) \times I\left[C(\mathbf{x}) \neq y \wedge S(\mathbf{x}) = y\right] \\
& - NT(\mathbf{x}) \times I\left[C(\mathbf{x}) \neq y \wedge S(\mathbf{x}) \neq y\right],
\end{aligned}
\tag{5}
$$

where $NT(\mathbf{x})$ and $NF(\mathbf{x})$ are the number of classifiers in S that classify $\mathbf{x}$ respectively correctly and wrongly.

3   Aim of This Work

A comparison between the effectiveness of directly using ensemble performance as the objective function, and using measures involving diversity, has been carried out by a few authors [1,12–15], often limited to the specific evaluation measure they were proposing, and using different and incomparable experimental setups (different data sets, base classifiers, ensemble construction methods, etc.). We also point out that only in [12,15] the use of pruning measures provided a statistically significant improvement over the use of ensemble performance.

Our aim is thus to carry out an extensive experimental investigation of FS/BS-based ensemble pruning methods, focused on the comparison between the use of ensemble performance as the objective function, and the use of measures involving diversity. To this aim, we focus on the basic FS/BS algorithm without back-fitting, and consider three kinds of objective functions:

1. Ensemble accuracy.
2. A generic diversity measure, focusing on the ones analyzed in [9]. Although diversity alone is deemed to be not effective for ensemble pruning [11,19], we consider also this option to provide a more direct evidence to these findings.
3. Pruning measures, which combine individual classifiers' performance and complementarity: we consider the ones described in Sect. 2, Eqs. (1)–(5).

We also consider another way to combine ensemble performance and diversity. Since diversity measures are not homogeneous to classification accuracy, to avoid combining them with individual classifiers' accuracy in an arbitrary way (e.g., by a linear combination), we use a two-stage FS/BS: first we select $M < N$ classifiers using either ensemble accuracy or diversity; then we further select $L < M$ classifiers using the other measure. Algorithm 2 shows the version in which ensemble accuracy is used at the first stage. In our experiments we considered both versions.

4   Experimental Setting

We chose 23 benchmark data sets from the UCI Machine Learning Repository Database,[2] with at least 350 samples, only numerical attributes, and without

---

[2] http://www.ics.uci.edu/~mlearn/MLRepository.html.

Algorithm 2. Two-stage Forward Selection algorithm for ensemble pruning

---
Input: a classifier ensemble E of size N; a desired ensemble size $L < N$; an intermediate ensemble size M, with $L < M < N$; a validation set V; a diversity measure d
Output: a subset of L classifiers from E

    step 1 (accuracy-based pruning): select from E an ensemble E' of size M using Algorithm 1, and using classification accuracy as the objective function m
    step 2 (diversity-based pruning): select from E' an ensemble S of size L using Algorithm 1, and using d as the objective function m
    return S

---

missing values (see Table 1). We used bagging to construct the original ensemble, majority voting as the combining rule, and two different base classifiers: multi-layer perceptron neural networks (MLP-NN) with one hidden layer containing ten units, and decision trees (DT). For MLP-NN we used the standard Matlab implementation[3], learning rate $\eta = 0.05$, and maximum number of training epochs equal to 300. For DTs we used the code of [11] (par. 2.A.2.1), with the Gini impurity criterion, χ^2 stopping criterion, and the default threshold equal to 1 for the pre-pruning stopping criterion. We set the size of the original ensemble to $N = 100$, and considered four different sizes of the pruned ensembles: $L = 5, 15, 25$ and 35.

We used only FS-based pruning. In the two-stage Algorithm 2 we set the size M of the first-stage pruned ensemble to $M = L + \lfloor (N - L)/2 \rfloor$. Since FS-based pruning starts from the best individual classifier, to better appreciate its effectiveness we chose the training set size of each data set in preliminary experiments, by maximizing the difference between the accuracy of an ensemble of 100 classifiers (constructed by bagging) and of the best individual classifier (see the right-most column of Table 1). We then set the size of the validation as one third of the training set, and used the remaining samples as a testing set. We also used only half of the validation set (one sixth of the training set) to evaluate the effect of validation set size on the performance of ensemble pruning. We considered the ten diversity measures analyzed in [9] (the ones in the top rows of Table 2), as well as measures in Eqs. (1)–(5), which combine into a single scalar the individual classifiers' performance and the complementarity between their errors (the ones in the bottom five rows of Table 2).

We carried out 20 runs of the experiments. At each run we selected the training, validation and testing sets by stratified random sampling (no data set was originally subdivided into a training and a testing set). We applied bagging to the training set, to construct the original ensemble of $N = 100$ classifiers. We then run Algorithm 1 separately using as the objective function the ensemble accuracy, each diversity measure, and the pruning measures in Eqs. (1)–(5). We also run the two-stage Algorithm 2 in both versions (using accuracy either at the first or at the second stage), for each diversity measure. We finally computed, separately for each data set, pruning method, base classifier,

---

[3] http://it.mathworks.com/help/nnet/ref/patternnet.html.

Table 1. Characteristics of the data sets. The two rightmost columns report the size of the training set for the two base classifiers, as a fraction of the whole data set.

Dataset	Samples	Classes	Features	Tr. set size	
				MLP-NN	DT
Australian	690	2	14	0.42	0.42
Balance scale	625	3	4	0.18	0.42
Blood transfusion	748	2	4	0.48	0.60
Breast cancer	699	2	9	0.30	0.12
Bupa	345	2	6	0.54	0.06
Checker board	1000	2	2	0.36	0.30
Coil 2000	9822	2	85	0.06	0.18
Cone tours	2000	3	2	0.06	0.24
Contraceptive	1473	3	9	0.36	0.60
ILPD	583	2	9	0.50	0.06
Laryngeal 2	692	2	16	0.06	0.48
Monk2	432	2	6	0.48	0.06
Page blocks	5473	5	10	0.06	0.42
Phoneme	5404	2	5	0.36	0.30
Pima Indians	768	2	8	0.54	0.30
Pop failures	540	2	20	0.42	0.30
Ring	7400	2	20	0.42	0.30
SaHeart	462	2	4	0.54	0.18
Sata log image seg	2310	7	19	0.44	0.30
Landsat Satellite	6435	7	36	0.60	0.48
Spam base	4601	2	57	0.42	0.30
Townorm	7400	2	20	0.12	0.30
Wine quality	4898	7	11	0.18	0.30

ensemble size L and validation set size, the average accuracy and its standard deviation on testing samples, over the 20 runs. Due to space limits, we make these results available only from our web site,[4] and only report the results of the statistical significance test. We compared the accuracy of pruned ensembles attained by Algorithm 1 using ensemble accuracy as the objective function, and using each of the other measures (both by Algorithms 1 and 2). To this aim we used the Wilcoxon signed-rank test, which is recommended in [5] for comparing two algorithms over multiple data sets. Our goal was to assess whether the difference was significant, and, if so, whether using ensemble accuracy as the objective function was the best or the worst option. Accordingly, we made two

---

[4] http://pralab.diee.unica.it/en/MCS2015Appendix1.

Table 2. Diversity measures (top ten rows, from [9]) and pruning measures (in the other rows, defined in Eqs. (1)–(5)) used in the experiments.

Diversity/pruning measure	Abbreviation
Entropy	E
Kohavi-Wolpert	KW
Coincidence failure diversity	CFD
Generalized diversity	GD
Interrater agreement	Kappa
Difficulty	Theta
Q Statistic	Q
Correlation	Rho
Disagreement	D
Double fault	DF
Uncertainty weighted accuracy	UWA
Partridge and yates' measure	PYM
Complementariness	Cs
Margin distance	MD
Concurrency	Cy

one-sided tests (at the $\alpha = 0.05$ level), evaluating the null hypotheses that FS-based pruning using ensemble accuracy (or a measure involving diversity) is not better than using a given measure involving diversity (or ensemble accuracy). Only if *both* null hypotheses are rejected, it can be concluded that there is no statistically significant difference between the two options.

5 Experimental Results

For each pruned ensemble size L, base classifier, and validation set size, Tables 3, 4, 5, 6, 7 and 8 report the comparison between FS-based pruning (Algorithm 1) using ensemble accuracy, and FS-based pruning implemented by Algorithm 1 using either a diversity or a pruning measure, and by Algorithm 2 combining ensemble accuracy and diversity.

Tables 3 and 4 clearly show that using ensemble accuracy often provides a better or comparable pruned ensemble than using any diversity measure alone, or a pruning measure. The only exceptions are GD (with $L = 15$) and UWA (with $L = 35$), using DT as the base classifier and a small validation set (see Table 3).

Interestingly, most of the cases when using diversity attained comparable results occur for three only measures: Entropy, Generalized Diversity and Kappa.

Tables 5, 6, 7 and 8, which refer to the two-stage FS algorithm combining ensemble performance and diversity, show a different pattern, instead. When a larger validation set is used, ensemble accuracy still produces often a better or

Table 3. Comparison of FS-based pruning (Algorithm 1) using ensemble accuracy vs. using each diversity or pruning measure, for different ensemble sizes L and validation set sizes. Base classifier: DT. 'A': using accuracy is statistically significantly better than using the corresponding diversity/other measures, over the 23 data sets; 'D': using the corresponding diversity/other measures is better than ensemble accuracy; '-': there is no statistically significant difference between the two measures.

Diversity measure	Ensemble size L							
	Val. size: 1/3 Tr. size				Val. size: 1/6 Tr. size			
	5	15	25	35	5	15	25	35
E	-	-	-	-	-	-	-	-
KW	A	A	A	A	A	A	A	A
CFD	A	A	A	A	A	A	A	A
GD	-	-	-	-	-	D	-	-
Kappa	-	-	-	-	-	-	-	-
Theta	-	A	-	A	A	-	-	-
Q	-	A	-	A	A	-	-	-
Rho	A	A	A	A	-	A	A	A
D	A	A	A	A	A	A	-	-
DF	A	A	A	A	A	A	A	A
UWA	A	A	A	A	-	-	-	D
PYM	-	-	-	-	-	-	-	-
Cs	A	A	A	A	A	A	A	A
MD	-	-	-	-	-	-	-	-
Cy	-	-	-	-	-	-	-	-

comparable pruned ensemble; however, for ensembles of DTs it never outperforms the combination of ensemble performance and diversity; moreover, it almost always performs worse with respect to the Double Fault (DF) measure. When a smaller validation set is used, together with DT classifiers, instead, combining ensemble accuracy and diversity is often better, or at least not worse, than using only ensemble accuracy (four right-most columns of Tables 5 and 7, vs the same columns of Table 3). Remarkably, this happens for most diversity measures.

These results seem to suggest that estimating the ensemble performance is the best option for FS-based pruning, provided that a sufficiently large validation set is available. Otherwise, a combination of ensemble performance and diversity can be advantageous, at least for some types of base classifiers. One possible explanation is that diversity measures have a *regularization* effect capable of preventing over-fitting, to some extent, as already argued in [12]. This is an interesting and non-straightforward property, which is worth investigating more throughly.

Table 4. Comparison of FS-based pruning (Algorithm 1) using ensemble accuracy vs. using each diversity or pruning measure, for a validation set size equal to 1/3 and 1/6 of the training set size. Base classifier: MLP-NN. See caption of Table 3 for the meaning of table entries.

Diversity measure	Ensemble size L							
	Val. size: 1/3 Tr. size				Val. size: 1/6 Tr. size			
	5	15	25	35	5	15	25	35
E	-	-	-	-	A	-	-	-
KW	A	A	A	A	-	-	-	-
CFD	A	A	A	A	-	-	-	-
GD	-	-	-	-	-	-	-	-
Kappa	-	-	-	-	-	-	-	-
Theta	A	A	A	A	-	-	-	-
Q	A	A	A	A	-	-	-	-
Rho	A	A	A	A	-	-	-	-
D	A	A	A	A	-	-	-	-
DF	A	A	A	A	-	-	-	-
UWA	-	-	-	-	-	-	-	-
PYM	-	-	-	-	-	-	-	-
Cs	A	A	A	A	A	A	A	A
MD	-	-	-	-	-	-	-	-
Cy	-	-	-	-	-	-	-	-

Table 5. Comparison of FS-based pruning (Algorithm 1) using ensemble accuracy vs Algorithm 2 using ensemble accuracy at the first stage and each diversity measure at the second stage. Base classifier: DT. See caption of Table 3 for the meaning of table entries.

Diversity measure	Ensemble size L							
	Val. size: 1/3 Tr. size				Val. size: 1/6 Tr. size			
	5	15	25	35	5	15	25	35
E	-	-	-	-	D	D	D	D
KW	-	-	-	-	D	D	D	D
CFD	-	-	-	D	D	D	D	D
GD	-	-	-	D	D	D	D	-
Kappa	-	-	-	-	D	D	D	-
Theta	-	-	-	-	D	D	D	-
Q	-	-	-	-	-	D	D	-
Rho	-	-	-	-	D	D	D	-
D	-	-	-	-	D	D	D	-
DF	-	D	D	D	D	D	D	D

Table 6. Comparison of FS-based pruning (Algorithm 1) using ensemble accuracy vs Algorithm 2 using ensemble accuracy at the first step and each diversity measure at the second stage, for a validation set size equal to 1/3 and 1/6 of the training set size. Base classifier: MLP-NN. See caption of Table 3 for the meaning of table entries.

Diversity measure	Ensemble size L							
	Val. size: 1/3 Tr. size				Val. size: 1/6 Tr. size			
	5	15	25	35	5	15	25	35
E	A	A	A	A	A	A	A	A
KW	A	A	A	A	A	A	A	A
CFD	-	-	-	-	A	-	D	-
GD	-	-	-	-	A	-	-	-
Kappa	A	A	-	-	A	A	A	A
Theta	A	-	-	-	A	-	-	-
Q	A	-	-	A	A	A	A	A
Rho	A	A	A	-	A	A	A	A
D	A	A	A	A	A	A	A	A
DF	D	D	D	D	-	-	-	-

Table 7. Comparison of FS-based pruning (Algorithm 1) using ensemble accuracy vs Algorithm 2 using each diversity measure at the first stage and ensemble accuracy at the second stage. Base classifier: DT. See caption of Table 3 for the meaning of table entries.

Diversity easure	Ensemble size L							
	Val. size: 1/3 Tr. size				Val. size: 1/6 Tr. size			
	5	15	25	35	5	15	25	35
E	-	-	-	-	D	D	D	-
KW	-	-	-	-	D	D	-	-
CFD	-	-	-	-	D	D	-	D
GD	-	-	D	D	-	D	D	D
Kappa	-	-	-	-	-	D	D	-
Theta	-	-	-	-	-	D	D	-
Q	-	-	-	-	-	D	-	-
Rho	-	-	-	-	D	D	D	-
D	-	-	-	-	D	D	-	-
DF	-	-	D	D	-	D	D	D

Table 8. Comparison of FS-based pruning (Algorithm 1) using ensemble accuracy vs Algorithm 2 using each diversity measure at the first stage and ensemble accuracy at the second stage, for a validation set size equal to 1/3 and 1/6 of the training set size. Base classifier: MLP-NN. See caption of Table 3 for the meaning of table entries.

Diversity Measure	Ensemble size L							
	Val. size: 1/3 Tr. size				Val. size: 1/6 Tr. size			
	5	15	25	35	5	15	25	35
E	-	A	A	A	-	-	-	-
KW	-	A	A	A	A	A	-	A
CFD	-	-	-	-	A	D	D	-
GD	-	D	-	-	-	D	D	-
Kappa	-	A	A	-	A	A	-	-
Theta	-	A	A	-	A	A	-	-
Q	A	A	A	A	A	-	A	A
Rho	-	A	A	A	A	A	-	A
D	A	A	A	A	A	-	-	A
DF	-	-	-	-	-	-	-	-

6 Discussion

We empirically investigated the effectiveness of explicitly using diversity measures for FS-based ensemble pruning, vs the simple estimation of ensemble accuracy. On the one hand, our results provide a more direct evidence in support of previous findings that using diversity measures alone is not effective for ensemble pruning [11,19], and in particular are in agreement with the well-established fact that diversity is not monotonically related to ensemble accuracy [11]. On the other hand, they suggest that, combined with the performance of individual classifiers, diversity can be useful to FS-based pruning when a small validation set is available. It seems therefore that diversity has a regularization effect. This possible effect has already been argued through the derivation of generalization bounds in [22], in the context of constructing ensembles of support vector machines, as well as in [12], in the context of FS-based ensemble pruning. However, in [12] the effect of different validation set sizes was not assessed, and only one diversity and two pruning measures were considered for comparison (Table 6).

To sum up, what our results provide is not a sharp conclusion either in favor or against the effectiveness of explicitly using diversity measures for ensemble pruning. Instead, and perhaps more interestingly, they provide some hints on the conditions under which diversity can be useful, and clearly suggest as a future research direction a more thorough investigation of the effect of validation set size. Our analysis can also be extended to other pruning methods categorized in [20] as optimization-based, which use genetic algorithms [7,23] or a kind of best-first search [18], where ensemble accuracy can also be used as the objective function. Finally, this investigation can be extended to regression problems, in which the exact Ambiguity decomposition includes a diversity term which does *not* depend on ground truth, contrary to most diversity measures for classification problems, including all the ones in [9] considered in this work, and the one in [3] derived from an *exact* Ambiguity-like decomposition; this allows it to be computed also on a set of *unlabeled* samples, thus potentially reducing the effect of over-fitting when a small set of (labelled) validation samples is available.

Acknowledgments. This work has been partly supported by the project CRP-59872 funded by Regione Autonoma della Sardegna, L.R. 7/2007, Bando 2012.

References

1. Banfield, R.E., Hall, L.O., Bowyer, K.W., Kegelmeyer, W.P.: EnsembleUWA diversity measures and their application to thinning. Inf. Fusion **6**(1), 49–62 (2005)
2. Brown, G., Wyatt, J.L., Harris, R., Yao, X.: Diversity creation methods: a survey and categorisation. Inf. Fusion **6**(1), 5–20 (2005)
3. Brown, G., Kuncheva, L.I.: "Good" and "Bad" diversity in majority vote ensembles. In: El Gayar, N., Kittler, J., Roli, F. (eds.) MCS 2010. LNCS, vol. 5997, pp. 124–133. Springer, Heidelberg (2010)
4. Caruana, R., Niculescu-Mizil, A., Crew, G., Ksikes, A.: Ensemble selection from libraries of models. In: 21st International Conference on Machine Learning, p. 18. ACM (2004)
5. Demsar, J.: Statistical comparisons of classifiers over multiple data sets. J. Mach. Learn. Res. **7**, 1–30 (2005)
6. Didaci, L., Fumera, G., Roli, F.: Diversity in classifier ensembles: fertile concept or dead end? In: Zhou, Z.-H., Roli, F., Kittler, J. (eds.) MCS 2013. LNCS, vol. 7872, pp. 37–48. Springer, Heidelberg (2013)
7. Ko, A.H.-R., Sabourin, R., de Souza Britto Jr., A.: Compound diversity functions for ensemble selection. Int. J. Patt. Rec. Artif. Int. **23**(4), 659–686 (2009)
8. Krogh, A., Vedelsby, J.: Neural network ensembles, cross validation, and active learning. In: Tesauro, G., Touretzky, D.S., Leen, T.K. (eds.) Advances in Neural Information Processing Systems 7, pp. 231–238. MIT Press, Cambridge (1995)
9. Kuncheva, L.I., Whitaker, C.J.: Measures of diversity in classifier ensembles and their relationship with the ensemble accuracy. Mach. Learn. **51**(2), 181–207 (2003)
10. Kuncheva, L.I.: A bound on kappa-error diagrams for analysis of classifier ensembles. IEEE Trans. Knowl. Data Eng. **25**(3), 494–501 (2013)
11. Kuncheva, L.I.: Combining Pattern Classifiers: Methods and Algorithms, 2nd edn. Wiley, Hoboken (2014)
12. Li, N., Yu, Y., Zhou, Z.-H.: Diversity regularized ensemble pruning. In: Flach, P.A., De Bie, T., Cristianini, N. (eds.) ECML PKDD 2012, Part I. LNCS, vol. 7523, pp. 330–345. Springer, Heidelberg (2012)
13. Margineantu, D.D., Dietterich, T.G.: Pruning adaptive boosting. In: 14th International Conference Machine Learning, pp. 378–387. Morgan Kaufmann (1997)
14. Martinez-Munoz, G., Suarez, A.: Aggregation ordering in bagging. In: International Conference on Artificial Intelligence and Applications, pp. 258–263 (2004)

15. Partalas, I., Tsoumakas, G., Vlahavas, I.P.: An ensemble uncertainty aware measure for directed hill climbing ensemble pruning. Mach. Learn. **81**, 257–282 (2010)
16. Partridge, D., Yates, W.B.: Engineering multiversion neural-net systems. Neural Comput. **8**(4), 869–893 (1996)
17. Prodromidis, A., Stolfo, S.J.: Pruning meta-classifiers in a distributed data mining system. In: Proceedings of the 1st National Conference on New Information Technologies, pp. 151–160 (1998)
18. Rokach, L.: Collective-agreement-based pruning of ensembles. Comp. Stat. Data Anal. **53**(4), 1015–1026 (2009)
19. Tang, E.K., Suganthan, P.N., Yao, X.: An analysis of diversity measures. Mach. Learn. **65**, 247–271 (2006)
20. Tsoumakas, G., Partalas, I., Vlahavas, I.: An ensemble pruning primer. In: Okun, Oleg, Valentini, Giorgio (eds.) Applications of Supervised and Unsupervised Ensemble Methods. SCI, vol. 245, pp. 1–13. Springer, Heidelberg (2009)
21. Ueda, N., Nakano, R.: Generalization error of ensemble estimators. In: International Conference on Neural Networks, pp. 90–95 (1996)
22. Yu, Y., Li, Y.-F., Zhou, Z.-H.: Diversity regularized machine. In: 22nd International Joint Conference on Artificial Intelligence, pp. 1603–1608 (2011)
23. Zhou, Z.-H., Wu, J., Tang, W.: Ensembling neural networks: many could be better than all. Artif. Intell. **137**(1–2), 239–263 (2002)
24. Yu, Y., Li, Y.-F., Zhou, Z.-H.: Diversity regularized machine. In: Proceedings of the 22nd International Joint Conference on Artificial Intelligence, pp. 1603–1608 (2011)
25. Zhou, Z.-H.: Ensemble Methods: Foundations and Algorithms. CRC Press, USA (2012)

Bio-Visual Fusion for Person-Independent Recognition of Pain Intensity

Markus Kächele[1], Philipp Werner[2], Ayoub Al-Hamadi[2], Günther Palm[1],
Steffen Walter[3], and Friedhelm Schwenker[1](✉)

[1] Institute of Neural Information Processing, Ulm University, Ulm, Germany
`{markus.kaechele,guenther.palm,friedhelm.schwenker}@uni-ulm.de`
[2] Institute of Information Technology, University of Magdeburg,
Magdeburg, Germany
`{philipp.werner,ayoub.al-hamadi}@ovgu.de`
[3] Department of Psychosomatic Medicine and Psychotherapy,
Ulm University, Ulm, Germany
`steffen.walter@uni-ulm.de`

Abstract. In this work, multi-modal fusion of video and biopotential signals is used to recognize pain in a person-independent scenario. For this purpose, participants were subjected to painful heat stimuli under controlled conditions. Subsequently, a multitude of features have been extracted from the available modalities. Experimental validation suggests that the cues that allow the successful recognition of pain are highly similar across different people and complementary in the analysed modalities to an extent that fusion methods are able to achieve an improvement over single modalities. Different fusion approaches (early, late, trainable) are compared on a large set of state-of-the art features for the biopotentials and video channels in multiple classification experiments.

1 Introduction

The nature of pain as a feedback mechanism to prevent harmful behaviour and to support self-preservation can be regarded a useful trait from an evolutionary point of view. However if the feedback characteristics miss their effect because a person is not able to act accordingly (e.g. somnolent patients, patients suffering from dementia), pain can be a heavy burden. Under certain circumstances, there is little correlation between subjectively experienced pain and tissue lesions or other pathological changes, the pain may even be completely unrelated. Therefore, the somatic pathology does not allow any conclusions to be drawn on subjectively experienced pain. In recent years, the focus of machine learning of human centered signals has witnessed a shift from the recognition of facial expressions and emotions to clinical fields of research such as depression [6], post-traumatic stress disorder [16] or pain. Research interest in automatic pain recognition has focused largely on recognition of facial expressions in painful situations for example using the UNBC-McMaster shoulder pain expression archive database [4,9] and only recently the in-depth investigation of biopotentials has led to encouraging findings [17] for automated pain recognition. Predictions based on both

© Springer International Publishing Switzerland 2015
F. Schwenker et al. (Eds.): MCS 2015, LNCS 9132, pp. 220–230, 2015.
DOI: 10.1007/978-3-319-20248-8_19

modalities combined as is it already very common in other sub-disciplines of machine learning is still an unexplored area and only very few works exist that leverage information fusion and multi expert systems [22]. This work aims to investigate fusion of biopotential and video data to improve the recognition of pain intensity in person-independent scenarios.

2 Dataset and Feature Extraction

In these experiments the BioVid Heat Pain database [19] is analysed. It comprises 90 participants ((1) 18-35 years (n = 30 years; 15 men, 15 women), (2) 36-50 years (n = 30; 15 men, 15 women), and (3) 51-65 years (n = 30; 15 men, 15 women)). The experimental setup consisted of a thermode that was used for the pain elicitation. The intensity was calibrated for each participant such that it divided the range between two reference levels (pain starts and pain is barely bearable) into 3 equally spaced intervals. Each of the 4 different stimulation strengths was applied 20 times to give rise to a total of 80 responses. During the experiments, high resolution video (from 3 different cameras), sensor data of a Kinect, and a biophysiological amplifier were recorded. The physiological channels included electromyography (EMG) (zygomaticus, corrugator and trapezius muscles), skin conductance level (SCL) and an electrocardiogram (ECG). The study was conducted in accordance with the ethical guidelines set out in the WMA Declaration of Helsinki (ethical committee approval was granted: 196/10-UBB/bal). The experiment was carried out twice. In the first round (part A), no facial EMG was attached to prevent deterioration of the video signal. In the second round (part B) a full set of EMG electrodes was attached (see Fig. 1).

2.1 Biophysiological Feature Extraction

Feature extraction for the biopotentials was performed after channel dependent pre-processing. The EMG and ECG channels were filtered using a Butterworth bandpass filter with the frequency ranges of [20, 250] Hz and [0.1, 250] Hz, respectively. This step was necessary to reduce noise and minimize the effects of trends in the signals. For the EMG signal an additional noise reduction procedure based on Empirical Mode Decomposition was applied [1]. For the EMG and SCL channel, a number of features based on signal amplitude and frequency such as peak height, peak difference, mean absolute difference, Fourier coefficients, bandwidth were computed as well as additional features based on entropy (approximate and sample entropy [14]), stationarity [3] and statistical moments. In the ECG signal, first the QRS complexes were detected then based on the differences between consecutive heart beats (RR intervals), the mean difference, the root mean sum of squared differences (RMSSD) as well as the slope of the regression line computed on the RR intervals, were calculated. All of the biopotential features were computed on a window of 5.5 s and the total number of features was 131.

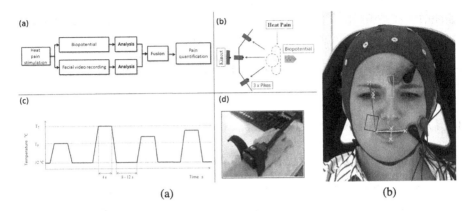

Fig. 1. Left: Experimental setting. The workflow of the experimental procedure and setup are depicted in the upper row. In the lower row, the different stimuli are depicted as well as a close up of the thermode. **Right**: Video Feature extraction: The depicted landmarks (yellow) are automatically detected and serve as input for the feature computation. Because of the attached EMG sensors, only the left part of the face is used (Color figure online).

2.2   Video Feature Extraction

From the video modality we extract facial expression and head pose features. The head pose is estimated from depth maps by fitting a generic head model to the measured point cloud [21]. This yields 3 rotation angles and 3 position parameters per frame. For facial expression features, we automatically detect facial landmarks on the mouth, right eye and right eyebrow with IntraFace [23]. Several distances between these landmarks measure facial deformation. These distances are measured in 3D [21] and include eye to brow distance, eye closure and mouth height, among others. Deepening of the nasolabial fold are measured through the mean gradient magnitude in the corresponding image region. Over time, each of these facial expression and head pose parameters yields a signal. For each signal we apply a low-pass filter and calculate the first and second temporal derivative of the resulting signal. Next, we extract statistical parameters of the low-pass filtered signal and its derivatives, namely the mean, median, standard deviation, range, inter-quartile range, inter-decile range and median absolute deviation. The extracted statistical parameters are used as facial expression and head pose features for the following analysis. The overall feature extraction method is detailed in [21,22]. For part B, only a subset of the features of [21,22] was used, as the EMG electrodes interfere with the landmark detection in the left part of the face.

3   Classification and Fusion Approaches

For the classification of the pain intensity levels, a number of classification and fusion techniques have been applied to the extracted features. The focus of this work is set on the applicability of early and decision fusion approaches for person-independent recognition of pain stimuli. Early fusion denotes the process of directly combining (i.e. concatenating) features before a classifier is trained. Late or decision fusion is applied after the individual classifiers have been trained by combining their predictions for unseen samples. Fixed mappings such as taking the mean or product of the individual predictions are common as well as trainable mappings such as the pseudo-inverse. Additionally the combination of early and decision fusion has been investigated by combining specific channels on feature level with subsequent fusion with other channels on decision level. In the literature (e.g. [8]) a variety of possible fusion mappings exist (grouped into fixed and trainable). As a fixed mapping we chose the sum rule and as a trainable mapping we decided for a pseudo-inverse trained on each of the classifiers' probability outputs per class. The classifiers of choice were an SVM with linear kernel (with softmax outputs) and a Random Forest. Since the dimensionality of the input data (especially when concatenated) reached several hundred, a feature selection algorithm has also been applied to reduce the set to the most discriminative ones. In the following, the individual components of the classification architecture are briefly introduced.

3.1   Support Vector Machine

A Support Vector Machine [18] is a classifier that maximizes the margin between the positive and negative classes. The optimal hyperplane $\mathbf{w}$ is achieved by optimizing

$$\min_{\mathbf{w},\xi} \frac{1}{2}\mathbf{w}^T\mathbf{w} + C\sum_i \xi_i \tag{1}$$

under the constraints $y_i(\mathbf{w}^T\mathbf{x_i} + b) \geq 1 - \xi_i, \forall i$ and $\xi_i \geq 0, \forall i$, where y_i is the label of sample $\mathbf{x_i}$. C is a parameter that controls the penalty of samples that lie inside the margin region (or on the wrong side of the hyperplane) for linearly non-separable problems. Optimization is commonly done by deriving the dual form of the optimization function and then using quadratic programming or the sequential minimal optimization (SMO) algorithm [12].

3.2   Random Forest

Random Forests [2] are ensembles of bagged decision trees that are trained on randomly drawn subsets of features. For each tree in each node, a split is computed based on an impurity measure such as the Gini index. A new sample is classified by querying each tree and a subsequent majority voting. Random Forests are very robust against changes in the input space and insensitive against parameter choices.

3.3   Pseudo-Inverse

A least-squares optimal linear mapping is obtained by computing the pseudo-inverse of the classifier outputs C^i for each classifier i and multiplying it with the desired values Y.

$$M^i = Y \lim_{\alpha \to 0_+} C_i^T (C_i C_i^T + \alpha I)^{-1} \tag{2}$$

This is commonly done using a hold out set. The mapping is then applied to the predicted outputs to obtain the final class memberships. For details, the reader is referred to [15].

3.4 Hybrid Sequential Floating Forward Selection

The sequential floating forward selection (SFFS) algorithm [13] is a so called wrapper method. An arbitrary classifier is used to evaluate the selected features while an enclosing routine optimizes which features to choose. The algorithm starts with the empty set and alternates the addition of a new (beneficial) feature and the subsequent removal of one or more features as long as the classification rate increases. Because all the features have to be added and tested individually, the process takes a long time if a huge number of features exist. Therefore a greedy pre-selection based on the minimum redundancy maximum relevance (mRMR) criterion [11] has been applied to filter promising feature candidates.

4   Experimental Validation

Different experiments have been carried out in order to investigate the discriminative power of each modality individually and also combined using different fusion techniques. Each of the following experiments has been carried out using a leave-one-subject-out cross validation to evaluate the generalization ability of the learning algorithms given unseen test subjects.

The first experiment was conducted with the setting of *no pain* (level 0) against the *upper pain threshold* (level 4) to test the feasibility of the task. As can be observed in Table 1, the recognition rate is notably higher than chance level, especially when considering the person-independent scenario. The best individual channel in this case is EMG (corrugator). Late fusion is able to improve the final accuracy by almost 4 %.

Table 1. Multimodal SVM classification. The results for part B underline the strong effect the late fusion has on the recognition rate. An improvement of almost 4 % over the best single modality can be observed.

Stimulus	EMG (zyg)	EMG (corr)	EMG (trap)	ECG	SCL	Video	Late (mean)	Early
0 vs. 4	0.678	0.728	0.613	0.635	0.670	0.719	**0.766**	0.682

To reduce the number of dimensions of the input features, additional feature selection has been applied. The effect of feature selection can be observed in Fig. 2. Applying the sequential floating forward selection (SFFS) [13] to each modality individually improves the results in some cases, however not in every case. Late fusion, on the other hand, benefits from the feature selection in both

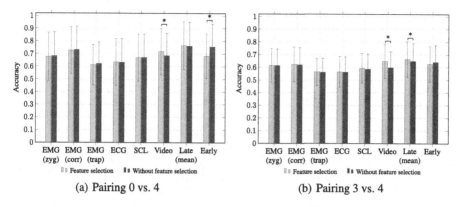

(a) Pairing 0 vs. 4   (b) Pairing 3 vs. 4

Fig. 2. Effect of feature selection. It can be seen that feature selection (red) improves the results in a few cases. However, late fusion consistently benefits from feature selection and is able to improve the already improved results even more. Early fusion on the other hand cannot benefit from the reduced feature set. Significance (indicated with asterisk) is computed using a Wilcoxon signed rank test with a significance level of 5 % (Color figure online).

depicted cases, improving the results over the fusion results without feature selection. Additionally, besides an improvement in accuracy, the amount of features is reduced by more than 90 % from 425 to only 41. It can be observed that early fusion becomes worse in this case, as the features were (greedily) selected for each channel individually.

Instead of using only late or early fusion, a combination of those schemes can be used on selected channels. For the next experiment, all the biopotentials are concatenated to a single channel and tested against the video channel. Classifying using a Random Forest yields some interesting results. While normally early fusion works better for the Random Forest, here the best result is obtained by late-fusing the two channels. For the comparison experiment where the channels are treated individually, early fusion works best with rates almost the same as the early rates in Fig. 3. For results on part B, the reader is referred to Fig. 4.

For this setting a variable importance estimate is generated by a permutation test on out-of-bag (OOB) samples of the Random Forest. This means that for each variable, the values are permuted across each sample that was not used for the training of a specific tree. If a variable is not important for the classification, the accuracy will not deteriorate. For important variables a drop in classification rate will be visible.

For each iteration in the cross validation (i.e. each person) the importance is computed for the case of early fusion of biophysiology and video. To obtain an estimate of features which are important for the whole classification process (i.e. not only for the person-specific setting), the individual importance estimates are ranked and the mean rank of each feature is calculated. The results show that the standard deviation of the stationarity for the skin conductivity is the most important feature, scoring an average rank of 1.0, which indicates that it

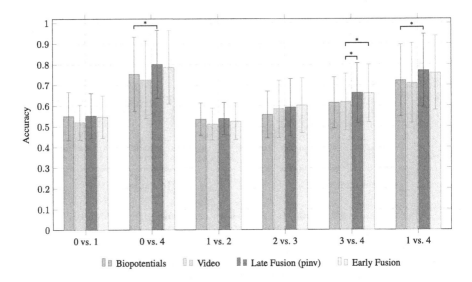

Fig. 3. Random Forest classification on part A. By early fusing the biopotentials, the decision fusion is able to improve over early fusion (which is normally superior when using RF) by 1 %–2 %. Significance (indicated with asterisk) is again computed using a Wilcoxon signed rank test with a significance level of 5 %. Note that significance is only indicated between fusion and the *best* single modality, other significant differences are omitted for clarity.

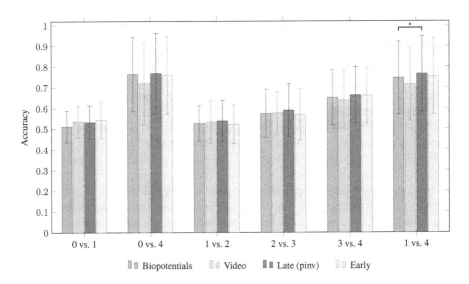

Fig. 4. Random Forest classification on part B. Decision fusion outperforms early fusion also on part B of the experiment when combining the biopotentials to a single channel. Significance (asterisk) by Wilcoxon signed rank test with a sign. level of 5 %. Again only indicated between fusion and the *best* single modality (see Fig. 3).

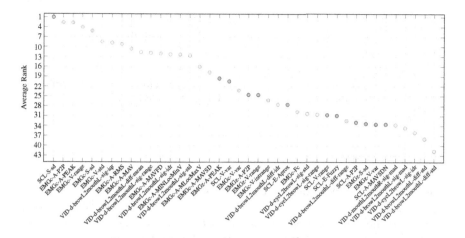

Fig. 5. Average rank in OOB importance estimation. As can be seen the top ranking feature (standard deviation of stationarity for SCL channel) was the most important one for each subject in the experiment. The different channels are indicated by the color as well as the prefix of the feature description. For details on the features, the reader is referred to [20]. For illustration purposes only the 40 best features are shown.

Table 2. Comparison of fusion approaches. In the setting without EMG it can again be observed that fusion outperforms the best individual channel. The fusion gains are about 4 % for both classifiers. For the SVM the pseudoinverse works best, while early fusion achieves the best results for the Random Forest.

Stimulus	ECG	SCL	Video	Late (mean)	Late (max)	Late (pinv)	Early
0 vs. 4 (SVM)	0.624	0.735	0.716	0.768	0.760	**0.772**	0.658
0 vs. 4 (RF)	0.581	0.744	0.727	0.774	0.778	0.771	**0.789**

is the most important feature for each subject in the experiment. The subsequent features are computed on the EMG channel and their ranks are also fairly consistent with 2.6471, 2.6824 and 4.0235, respectively (Fig. 5).

To further investigate the effect of fusion, an experiment was conducted using only the ECG and SCL signals together with video. Different late fusion techniques are tested with both classifiers on part A of the data collection. Table 2 summarizes the results. It can be seen that fusion improves over the best single modality in both cases.

Analysis of the other class pairings of the different pain levels indicates that it becomes more challenging the closer the intensity of the different pain stimuli (compare Figs. 3 and 4).

In summary it looks like the Random Forest has a slightly better performance on this data set. This can be attributed (in part) to its more robust nature and parameter insensitivity. Tuning of an SVM on the other hand involves the choice of a suitable kernel (with parameter) and regularization parameter.

A grid search has been conducted but the selection of different parameters might yield an improvement over the presented results (especially in the setting with feature selection).

While the late fusion mappings perform similarly, the trainable combiner in the form of a pseudoinverse seems to slightly outperform the other mappings. It should however be noted that early fusion normally works better in combination with a Random Forest.

5 Conclusion

In this work, fusion mechanisms for the person-independent recognition of levels of pain have been investigated. The experimental validation showed that both, the video channel and the biopotentials allow discrimination of pain from unseen persons. However in combination, the two modalities have the highest discrimination rate, which suggests that the two channels capture different cues and complement each other.

Avenues for future work include the investigation of heterogeneous time windows as each channel individually has another resolution of interest. Other possibilities include the systematic analysis of which channels to combine at what level, as it has been shown in related fields of work (affective computing: [5,7]), that it can be beneficial to explore the space of possible fusion architectures. For future recordings, the addition of an audio channel can be beneficial because reactions to a pain stimulus can also invoke paralinguistic expressions. The recordings can then be analysed as an additional modality using techniques from the field of affective computing [10].

Acknowledgments. This paper is based on work done within the Transregional Collaborative Research Centre SFB/TRR 62 *Companion-Technology for Cognitive Technical Systems* funded by the German Research Foundation (DFG). Markus Kächele is supported by a scholarship of the Landesgraduiertenförderung Baden-Württemberg at Ulm University.

This work was performed on the computational resource bwUniCluster funded by the Ministry of Science, Research and Arts and the Universities of the State of Baden-Württemberg, Germany, within the framework program bwHPC.

References

1. Andrade, A.O., Kyberd, P., Nasuto, S.J.: The application of the Hilbert spectrum to the analysis of electromyographic signals. Inf. Sci. **178**(9), 2176–2193 (2008)
2. Breiman, L.: Random forests. Mach. Learn. **45**(1), 5–32 (2001)
3. Cao, C., Slobounov, S.: Application of a novel measure of EEG non-stationarity as 'Shannon-entropy of the peak frequency shifting' for detecting residual abnormalities in concussed individuals. Clin. Neurophysiol. Official J. Int. Fed. Clin. Neurophysiol. **122**(7), 1314–1321 (2011)

4. Hammal, Z., Cohn, J.F.: Automatic detection of pain intensity. In: Proceedings of the 14th ACM International Conference on Multimodal Interaction, ICMI 2012, pp. 47–52. ACM, New York (2012)
5. Kächele, M., Glodek, M., Zharkov, D., Meudt, S., Schwenker, F.: Fusion of audio-visual features using hierarchical classifier systems for the recognition of affective states and the state of depression. In: De Marsico, M., Tabbone, A., Fred, A. (eds.) Proceedings of the International Conference on Pattern Recognition Applications and Methods (ICPRAM), pp. 671–678. SciTePress (2014)
6. Kächele, M., Schels, M., Schwenker, F.: Inferring depression and affect from application dependent meta knowledge. In: Proceedings of the 4th International Workshop on Audio/Visual Emotion Challenge, AVEC 2014, pp. 41–48. ACM, New York (2014)
7. Kächele, M., Schwenker, F.: Cascaded fusion of dynamic, spatial, and textural feature sets for person-independent facial emotion recognition. In: Proceedings of the International Conference on Pattern Recognition (ICPR), pp. 4660–4665 (2014)
8. Kuncheva, L.: Combining Pattern Classifiers: Methods and Algorithms. Wiley, Hoboken (2004)
9. Lucey, P., Cohn, J.F., Prkachin, K.M., Solomon, P.E., Matthews, I.: Painful data: the UNBC-McMaster shoulder pain expression archive database. Image Vis. Comput. J. **30**, 197–205 (2012)
10. Meudt, S., Zharkov, D., Kächele, M., Schwenker, F.: Multi classifier systems and forward backward feature selection algorithms to classify emotional coloured speech. In: Proceedings of the International Conference on Multimodal Interaction, ICMI 2013, pp. 551–556. ACM, New York (2013)
11. Peng, H., Long, F., Ding, C.: Feature selection based on mutual information criteria of max-dependency, max-relevance, and min-redundancy. IEEE Trans. Pattern Anal. Mach. Intell. **27**(8), 1226–1238 (2005)
12. Platt, J.C.: Fast Training of Support Vector Machines Using Sequential Minimal Optimization, pp. 185–208. MIT Press, Cambridge (1999)
13. Pudil, P., Novovičová, J., Kittler, J.: Floating search methods in feature selection. Pattern Recogn. Lett. **15**(11), 1119–1125 (1994)
14. Richman, J.S., Moorman, J.R.: Physiological time-series analysis using approximate entropy and sample entropy. Cardiovasc. Res. **278**(6), 2039–2049 (2000)
15. Schwenker, F., Dietrich, C.R., Thiel, C., Palm, G.: Learning of decision fusion mappings for pattern recognition. Int. J. Artif. Intell. Mach. Learn. (AIML) **6**, 17–21 (2006)
16. Stratou, G., Scherer, S., Gratch, J., Morency, L.P.: Automatic nonverbal behavior indicators of depression and PTSD: exploring gender differences. In: Humaine Association Conference on Affective Computing and Intelligent Interaction (ACII), pp. 147–152 (2013)
17. Treister, R., Kliger, M., Zuckerman, G., Aryeh, I.G., Eisenberg, E.: Differentiating between heat pain intensities: the combined effect of multiple autonomic parameters. Pain **153**(9), 1807–1814 (2012)
18. Vapnik, V.N.: Statistical Learning Theory. Wiley, New York (1998)
19. Walter, S., Gruss, S., Ehleiter, H., Tan, J., Traue, H., Werner, P., Al-Hamadi, A., Crawcour, S., Andrade, A., Moreira da Silva, G.: The BioVid heat pain database data for the advancement and systematic validation of an automated pain recognition system. In: 2013 IEEE International Conference on Cybernetics (CYBCONF), pp. 128–131, June 2013

20. Walter, S., Gruss, S., Limbrecht-Ecklundt, K., Traue, H.C., Werner, P., Al-Hamadi, A., Diniz, N., da Silva, G.M., Andrade, A.O.: Automatic pain quantification using autonomic parameters. Psychol. Neurosci. **7**, 363–380 (2014)
21. Werner, P., Al-Hamadi, A., Niese, R., Walter, S., Gruss, S., Traue, H.C.: Towards pain monitoring: facial expression, head pose, a new database, an automatic system and remaining challenges. In: Proceedings of the British Machine Vision Conference, pp. 119.1–119.13. BMVA Press (2013)
22. Werner, P., Al-Hamadi, A., Niese, R., Walter, S., Gruss, S., Traue, H.C.: Automatic pain recognition from video and biomedical signals. In: International Conference on Pattern Recognition, pp. 4582–4587 (2014)
23. Xiong, X., De la Torre, F.: Supervised descent method and its applications to face alignment. In: 2013 IEEE Conference on Computer Vision and Pattern Recognition (CVPR), pp. 532–539 (2013)

Author Index

Printed in the United States
By Bookmasters